GPRS Networks

GPRS Networks

**Geoff Sanders, Lionel Thorens, Manfred Reisky,
Oliver Rulik and Stefan Deylitz**

All of

tfk GmbH, Germany

WILEY

Other Wiley Editorial Offices

John Wiley & Sons Inc., 111 River Street, Hoboken, NJ 07030, USA

Jossey-Bass, 989 Market Street, San Francisco, CA 94103-1741, USA

Wiley-VCH Verlag GmbH, Boschstr. 12, D-69469 Weinheim, Germany

John Wiley & Sons Australia Ltd, 33 Park Road, Milton, Queensland 4064, Australia

John Wiley & Sons (Asia) Pte Ltd, 2 Clementi Loop #02-01, Jin Xing Distripark, Singapore 129809

John Wiley & Sons Canada Ltd, 22 Worcester Road, Etobicoke, Ontario, Canada M9W 1L1

Wiley also publishes its books in a variety of electronic formats. Some content that appears
in print may not be available in electronic books.

British Library Cataloguing in Publication Data

A catalogue record for this book is available from the British Library

ISBN 0-470-85317-4

Typeset in 10/12pt Times by Laserwords Private Limited, Chennai, India
Printed and bound in Great Britain by TJ International, Padstow, Cornwall
This book is printed on acid-free paper responsibly manufactured from sustainable forestry
in which at least two trees are planted for each one used for paper production.

Contents

Preface

The aim of this book is to provide the reader with a thorough grounding in the General Packet Radio Service – GPRS. The authors have assumed that the reader will have at least a basic understanding of GSM and the associated theory and terminology. The introduction contains a basic review of GSM to ensure that the reader is clear on the main aspects of circuit switched technology, as this will make the differences and advantages of packet oriented GPRS both more apparent and easier to understand.

The book will endeavour to take the reader beyond the specifications and into other crucial areas such as applications, implementation and other near-future aspects such as E-GPRS, the migration towards all-IP networks and the role played by GPRS in enabling packet data transfer from UMTS.

Wherever possible real-world examples will be given to support and illustrate the text. Readers of *GPRS Networks* have the following options with regard to obtaining further information on GPRS and other telecommunications subjects.

- for questions arising from the book please write a brief email to gprs@tfk.de where the authors and technical staff will be pleased to be of assistance
- to obtain .pdf format colour images of the graphics and diagrams in the book, please visit the Wiley ftp server at ftp://ftp/pub/books/sanders

To recommend the book to a friend or colleague please send them the following link www.gprsbook.tfk.de
The authors would very much appreciate your feedback about the book. Please send all comments to gprs@tfk.de

Introduction

Today no one doubts that the introduction of GSM represented a revolution in mobile communications. As this book goes to press, there are more than 773 million GSM users worldwide and the number is growing by the minute. This revolution, which has taken place over the last 12 years, has had an enormous impact on our daily lives. Mobile telephony has changed the way we communicate and the way we do business. Friends and colleagues can be reached almost anywhere at any time, we are no longer tied to desk-top machines for e-mails and faxes, and the mobiles themselves are replacing not only telephones but also dictaphones, personal organizers and phone books.

'Convergence' has been the buzzword in recent years. The above is just the beginning and GPRS is the key. As multifunction mobile devices replace cellphones, PDAs and notebook PCs, there will be an ever increasing demand for bandwidth. Home Internet users are used to the level of service offered by 64 kbits^{-1} ISDN or the higher rates offered by the various DSL services. These users will increasingly demand the same level of service from their mobiles, and as the volume of users goes up, content and service providers will respond and the network operators will have to provide the capacity.

Originally, 3G was seen as the means to satisfy the demand for bandwidth and GPRS was viewed by many simply as a stepping stone on the way from narrowband GSM to wideband UMTS. However, the worldwide delay in 3G rollout has led to GPRS being seen in a different light, particularly with the development of E-GPRS. The promise of low-cost reliable mobile Internet access has led to many operators implementing GPRS, but, whereas GSM was seen as a communications technology, GPRS is seen more as an enabling technology. The ability to run an application on a mobile device that can send and receive packet data at high speed makes possible such things as mobile Internet, remote device control, multi-player gaming, location-dependent information services, and m-commerce.

GPRS will continue to change the way we live, just as GSM has already done.

1

Mobile Radio Evolution

1.1 Trend from Speech to Data Transmission

Beginning with the first single cell mobile telephone services in the 1940s and continuing through the first and second generations of mobile telephone services, the primary function of the mobile device was to enable speech calls to be set up between the mobile user and a fixed network of base stations (BTS) and telephone exchanges.

This scenario has changed dramatically in recent years. With the introduction of first SMS and then WAP, few people now support the idea that a mobile device is only for speech calls, and just as the rate of change in mobile applications is accelerating, so is the rate of acceptance. As more and more people make use of mobile services, the multifunction portable device is becoming ever more indispensable in our daily lives.

This is an enormous departure from accepted norms. The trend from mobile speech to data is gaining acceptance faster than any other technology ever invented. The Internet took decades to reach its present form, and only since the mid-1990s has it enjoyed such widespread utilization from home users. GSM however, in just 10 short years, achieved over 80 % market penetration in some European countries! Compared with other industries, the rate of change in mobile technology is unprecedented. The automobile industry introduces a new model in about 5 years; the fashion industry also measures change in terms of years. Even the computer industry, which has experienced quantum leaps in terms of hardware technology and applications software, has not had the kind of acceptance which mobile telecommunications are having.

The fact of the matter is that all these industries – one could almost say *all* industries – are going to be directly affected by the trend towards applications based on mobile data transfer. It is already possible to send an SMS during a normal speech call, download information from the Internet via WAP, and access web sites and e-mail via a notebook or PDA connected to a mobile phone. This is only the beginning: it is predicted that by 2010 the volume of new mobile phone sales in Europe and North America will start to level off. This is the first indicator of the coming boom in mobile services based on packet data transmission. Already some manufacturers have started to develop and market solutions for m-services and m-commerce as they are already aware that future revenues will be

GPRS Networks.   G. Sanders, L. Thorens, M. Reisky, O. Rulik, S. Deylitz
© 2003 John Wiley & Sons, Ltd.   ISBN: 0-470-85317-4

generated not through equipment sales, but by providing the customer with new ways to make use of that equipment.

We can already see the start of this trend. The SIM tool kit introduces extra functionality onto the SIM card, which allows the user to subscribe to various service providers who use the tool kit to run their own proprietary software. This represents a major step forward in that it is no longer necessary to modify the mobile device or the SIM card in order to introduce new or improved services. This in turn means that the user can take advantage of new services at any time without having to invest in new equipment.

The automobile industry is a good example of current trends. Almost from the beginning of GSM, manufacturers at the top end of the market have been installing mobile phones in their cars. More recently, navigation systems based on GPS satellite positioning systems combined with electronic maps stored on CD-ROMs have become standard items on such vehicles. In recent years, these two systems have been linked together (converged) so that the navigation system receives traffic reports via SMS and uses this data to avoid traffic jams. A further enhancement is to store such things as hotel and tourist information on the navigation CD. Several manufacturers have now moved on to using DVDs as a storage medium due to the limited storage capacity of a CD-ROM.

What is the next step? The answer is to do away with on-board storage completely and use the general packet radio service (GPRS) to download the required information as and when required. Once this step has been taken, the on-board systems only need to be slightly upgraded to enable full Internet access. Such systems are in fact currently available as custom solutions, so it is only a matter of time before they become options/standard features offered by the manufacturer. This step is seen by many as essential to the continued evolution of in-car systems. The problem with CD/DVD systems is that the data is out of date almost as soon as it is released and cannot easily be updated. Further, such storage media offer a very limited amount of data as far as hotel and tourist information is concerned, added to which price and availability details cannot be included for the reason stated previously. A GPRS based system, which accesses a centralized server network, would always have access to the latest traffic information and could provide web links to relevant hotel, restaurant and tourist information web sites, based on GPS location information provided via the mobile.

As can be seen from the above example, there is a general trend towards increasing convergence of previously independent systems. In this we can also observe a trend in the way the Internet is being utilized. Previously, servers attached to the World Wide Web (www) enabled information exchange across the net. Later, e-commerce changed the face of the Web as it evolved from being purely an 'Information superhighway' (data, e-mail, gaming) to a way of doing business on a global scale. Now we see a further trend towards centralized, Web-based, servers replacing discrete/portable storage media. Such servers already provide business users with storage for their documents, presentations and other files. They can be accessed/updated online, avoiding the risk of having important information corrupted or stolen from a portable device. Such systems could replace in-car data storage – or any other application that depends on information stored on CD or DVD – using GPRS to give mobile access to the Internet.

The trends towards applications which can use GPRS are plainly visible. The success of i-mode in Japan and its subsequent release in Europe has demonstrated that there is a significant demand for mobile services based on high-speed packet data transfer. In its first year of business, i-mode attracted over 20 million users in Japan. If this success is repeated in Europe (where GPRS is the bearer) it will not only prove the technology and the business model, but will also provide a driving force for new services based on the general packet radio service.

1.2   The Third Generation

Among the many questions asked about 3G, one of the most frequent is 'Why do we need it?'. The answer lies in the planning. 2G was a response to a need for cheaper, more reliable, mobile communications which could easily cross international boundaries. The technology evolved through different phases which can be roughly characterized as: speech $\rightarrow$ speech related services $\rightarrow$ data services. Mobile Internet access was not initially envisaged, and this is why GSM has had to be supplemented with such things as HSCSD, GPRS and EDGE in the final phase (phase 2+). 3G, on the other hand, was intended from the outset to enable high bit-rate mobile data services (in addition to speech calls and traditional services) and has been planned accordingly.

The evolution towards 3G and beyond is happening on different levels. On the technological level we have a new radio access network, the UTRAN (UMTS radio access network), which interfaces with the GSM circuit switched (CS) core network for speech calls and the GPRS packet oriented (PO) core network for data transfer. On the services level, GSM required the addition of first the intelligent networks (IN) platform, to enable a greater variety of national services, and then the CAMEL (customized applications for mobile network enhanced logic) platform to make these services available while roaming outside the home PLMN. Modifications were needed to the radio access technology to enable high speed data services to be implemented. UMTS has been designed so that the services provided to the end-user are independent of the radio access technology.

On the network level, the current evolutionary trend is towards so called 'all-IP' networks. The convergence of CS and PO technologies already begun with the implementation of GPRS will reach its logical conclusion at some point in the future when the radio access networks, GERAN (GSM/EDGE radio access network) and UTRAN, will connect directly to a packet data network. Speech will be carried as voice over IP (VoIP) and will be routed from the source to the destination in the same way that other data packets are routed from one application to another.

The technological evolution from FDMA in 1G, through the combination of FDMA and TDMA in 2G, and on to CDMA in 3G, is a clear progression with each new development seeking to solve the problems of the previous generation. The evolution of the services offered to the end-user, however, is not such a linear process and thus a clear development path is difficult to identify. Providers are using the Internet as a base to offer their content and services to end-users with a great variety of terminal equipment. To maximize their revenues, they need to reach the maximum possible number of people and so must serve users with equipment from different phases and different generations. SMS, HSCSD, GPRS and WAP can all be used as means to bring value added services to the user and here is where the real strength of GPRS can be seen. The GPRS infrastructure can be utilized to deliver SMS and WAP services to both GSM and UMTS users. Here again we see evolution resulting in convergence. The implementation of GPRS was originally seen as a bridge between 2G and 3G but in actual fact has resulted in developments that have a direct affect on both generations.

1.3   GSM – The Global System for Mobile Communications

GSM is currently going through a period of optimization that is likely to continue for some time, as developers refine and enhance the system to get the best possible performance from existing GSM networks. Many of these refinements directly or indirectly affect GPRS and so will be described in the following sections.

1.3.1   Reasons for Success

Originally GSM stood for 'groupe speciale mobile' and was intended to be a new telecommunications standard for Europe. The success of the standard meant that the name had to be changed to reflect the truly worldwide application of GSM and became the 'global system for mobile communications'. The basis of GSM's success can be summarized as follows:

- an open standard – anyone can have access to the specifications,
- standardized interfaces – multi-vendor solutions possible,
- designed with roaming as a prerequisite – network architecture and procedures,
- encrypted transmission – ensures privacy/security of speech/data,
- subscriber identity module (SIM) – enables effective subscriber control,
- efficient use of available frequency spectrum – FDMA and TDMA,
- low power requirement – small battery size also means small mobiles,
- digital transmission – gives high speech quality and enables signalling for services,
- evolutionary implementation concept – upgradeable with downwards compatibility.

1.3.2   The GSM Air Interface

It is important to have an appreciation of the air interface Um (unlimited mobility) as GPRS shares this interface with GSM. In earlier implementations, GSM and GPRS channels were kept separate – i.e. separate frequencies were reserved for each one. With later releases, GSM and GPRS traffic channels can be mixed on the same frequency. The following description covers the main structure and terminology of the Um – signalling, protocols and the allocation of traffic channels will be covered in detail later.

The efficient use of the available radio spectrum was a key issue in the development of GSM. The uplink (UL) and downlink (DL) frequency blocks are divided into frequency bands with a bandwidth of 200 kHz. Each frequency band is divided in the time domain to give eight separate channels called 'time slots'. Hence, up to eight users can share the same frequency as their individual transmissions are separated in time – this is called time division multiple access (TDMA). The eight time slots on a single frequency band are called a 'TDMA frame'.

When a traffic channel (TCH) has been set up – i.e. the mobile station (MS) is assigned a frequency band and an UL/DL time slot number (the grey shaded time slots in Figure 1.1 represent the UL and DL time slots for one subscriber) – the MS can send and receive digitized speech, signalling messages, or data via this channel. In the case of speech, the analog voice signal must first be digitized and then channel coded (compressed + redundancy added) before it can be transmitted. Three coding schemes have been defined for this purpose: full rate (FR), enhanced full rate (EFR) and half rate (HR). FR and EFR both transmit in consecutive TDMA frames but HR gets its name from the fact that it

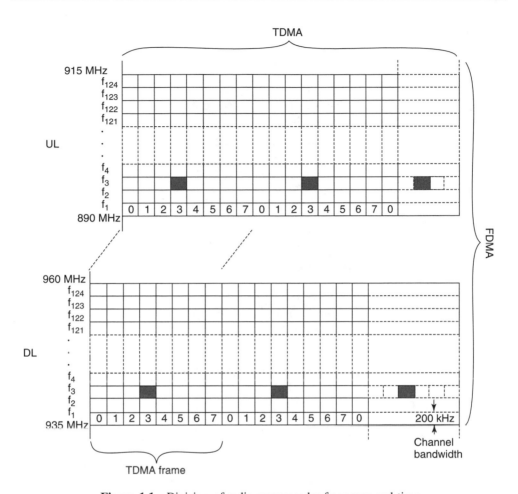

Figure 1.1   Division of radio resources by frequency and time

only makes use of a time slot in every second frame. This means that two mobiles using HR can share the same time slot (Figure 1.2).

The need to implement flexible allocation of GSM and GPRS channels on the Um has led to a refinement in the BSS software called active HR pairing. As calls start and finish using HR, we can be left with a situation where we have no channels available for GPRS because they are all occupied by single (unpaired) HR calls. A system is therefore required to force intracell handovers so that two unpaired HR calls can be moved onto one channel. Active HR pairing ensures that HR calls occupy the minimum number of time slots so that channels for FR speech connections and GPRS are not blocked. This also reduces co-channel interference as fewer frequencies are in use (Figure 1.3).

It was mentioned above that time slots can carry speech, signalling, or data – i.e. a time slot is just a container – but it is important that we understand how this works as there are some differences between GSM and GPRS.

If the MS is moving around during a phone call, then the distance to the BTS is changing. This means that the propagation delay (the 'travel time' for the time slot)

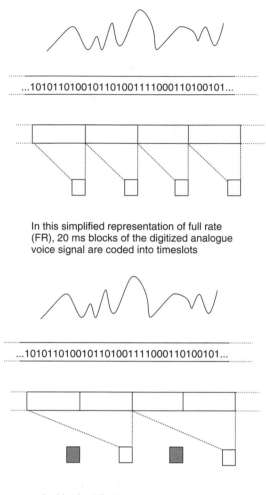

In this simplified representation of full rate
(FR), 20 ms blocks of the digitized analogue
voice signal are coded into timeslots

In this simplified representation of half rate
(HR), the HR coding scheme only uses every
second time slot. The shaded timeslots can be
used by another mobile

Figure 1.2   Full rate and Half rate coding

between the MS and BTS, as well as the received signal strength, will also change. To
ensure that the signal arrives at the BTS at the right time with the right signal strength, the
BTS sends timing advance (TA) and power control (PC) values to the MS via signalling
messages, roughly twice every second. These signalling messages are sent in the same
time slot used for the coded speech information, but not simultaneously. For FR, the BTS
sends speech in the time slot for the first 12 TDMA frames, but the thirteenth frame is
used to send a time slot containing signalling for the MS – e.g. PC and TA. Then 12
more speech time slots are sent, followed by an idle slot containing nothing. This gives
us the timing structures shown in Figure 1.4.

X = BTS signalling channel
G = GPRS channel
H = Half rate voice channel
h = half rate voice channel

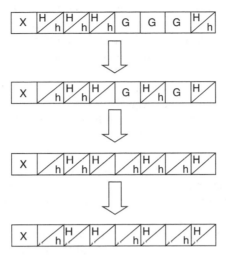

In this situation, there are no time slots availble for GPRS or full rate calls.

Active HR pairing resolves the problem by using intracell handovers to pair up HR calls in single timeslots.

The following pairing operation makes three time slots available for GPRS or FR calls.

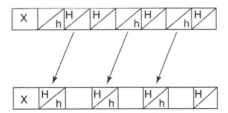

Figure 1.3 Active HR pairing

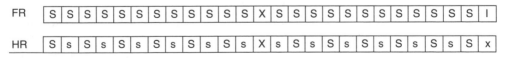

For FR: S = speech X = signalling I = idle
For HR: S = speech for first mobile X = signalling for first mobile s = speech for second mobile
X = signalling for second mobile

Figure 1.4 Multi-frames for voice traffic channels

As you can see, the idle time slot is necessary because it is used for the signalling to the second MS during a HR call. These structures are called 'multi-frames' and are made up of 26 time slots (each time slot belongs to one TDMA frame). GPRS uses a 52 time slot multi-frame for traffic channels which will be described later in the appropriate section.

In addition to the traffic channel multi-frame there are also signalling multi-frames. Just as the traffic channels (TCH) need to have a structure so that the MS can send and receive different kinds of information, the broadcast and control channels also need a structure to enable registration, authentication and call set-up messages to be exchanged between the BTS and several mobiles. The UL and DL multi-frames have different structures but each is based on a sequence of 51 TDMA frames.

At this point it is important to be clear about the different kinds of channels. A time slot is known as a 'physical' channel as it can be clearly defined in time and space. The physical channel assigned to an MS can be used to carry different information flows – i.e. speech, signalling or data. These information flows can be thought of as 'logical' channels being carried at different times by the physical channel. The multi-frame defined for signalling between base station and mobile ('X' in Figure 1.3.) represents a single physical channel carrying many different logical channels in a pre-defined sequence which repeats every 51 TDMA frames.

In earlier releases, GPRS makes use of the same logical channels as GSM in the signalling multi-frame. Later releases introduce GPRS specific channels. For example, the first signalling message from the MS to the BTS is sent on the random access channel (RACH), later releases of GPRS implement a PRACH – packet random access channel.

Depending on the number of carriers supported by the BTS there can be one or more time slots reserved for such signalling. All BTS have one carrier configured to transmit at full power. The broadcast channel (BCCH) is transmitted on this carrier in the first DL time slot (called 'TS0' as time slots are numbered from 0 to 7). For a BTS with only two carriers, TS0 on the UL and DL will be used to receive and send all the logical signalling channels. As more carriers are added, more time slots must be reserved for signalling – i.e. more users = higher signalling load = more signalling channels required (Figure 1.5). This impacts GPRS directly and will be covered in the GPRS signalling section.

Until now the time slots have been treated as an abstract concept – i.e. subdivisions of a carrier frequency – but as we will later be looking at a new modulation method (EDGE), it is necessary to understand how GSM actually gets the bits across the air interface.

Time slots are transmitted in the form of a short duration (577 µs) radio signal called a 'burst'. Based on the power and timing factors (PC and TA) received from the BTS, the MS ramps up its output to the given PC value, transmits the modulated bits with the given TA and ramps down again (see Figure 1.6).

The bits are modulated using Gaussian minimum shift keying (GMSK), a low-noise method which makes efficient use of the available 200 kHz channel bandwidth. GMSK shifts between two frequencies, each 67.7 kHz above or below the assigned carrier frequency for the channel – the absolute radio frequency carrier number (ARFCN). Simply switching between the two frequencies would produce a lot of noise (splash) so phase shifting is used both to generate the frequencies and to move smoothly between them by varying the rate of phase change. GMSK transmits '1' bits using one frequency and '0' bits on the other frequency.

The following represents a base station with two
carriers, i.e. 2UL and 2DL

X contains all BTS to MS signalling channels

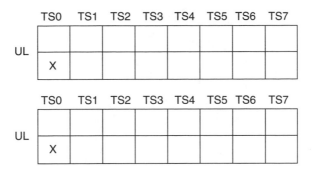

The following represents a base station with three
carriers, i.e. 3UL and 3DL

X_1 contains the broadcast channel (BCCH)
X_2 contains dedicated control channels

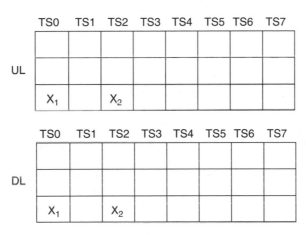

Figure 1.5 Time slots reserved for signalling channels

The GMSK approach used by GSM transmits the bit stream one bit at a time. EDGE
however, uses 8PSK which transmits three bits at a time. Combining EDGE with GPRS
(E-GPRS), bit rates approaching half a megabit per second are theoretically possible.

1.3.3 GERAN and Cell Planning

As GPRS makes use of the existing GSM radio access network it is important to understand how the cell/frequency planning is performed and how this can be optimized to
give better service to both speech and data users. The following is a brief review of
the GERAN.

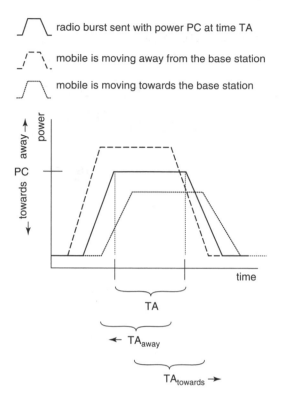

Figure 1.6   Radio burst characteristics as MS moves towards/away from the BTS

As far as the cellular structure of a GSM mobile network (PLMN) is concerned, we are used to seeing neat honeycomb diagrams of hexagonal cells. On a theoretical level the cells are round and have regular overlaps where handovers can take place. In actual fact the shape of the cells are greatly affected by screening, multi-path propagation and interference effects and are not the tidy circles we would like them to be. Such effects must be considered by the planners. Propagation models are developed to help plan the radio network. Reflection results in attenuation of the signal and can lead to areas with no coverage – so called 'dark spots' – as can screening by buildings. External sources can also attenuate or distort the radio signals but one of the biggest challenges facing radio planners is in optimizing the frequency reuse patterns (as follows). Standard GSM900 has 124 frequency channels and the planner's job is to ensure that neighbouring cells do not use the same frequencies as they will interfere with each other. There are various planning models in use, each of which defines a pattern of cells with different frequencies in each cell. The pattern – called a 'cluster' – can be repeated to ensure that the distance between cells with the same frequency is large enough to minimize interference – this is the so called 'frequency reuse distance' (Figure 1.7).

Until the advent of synchronized frequency hopping, the minimum cluster size was seven cells as shown in Figure 1.7. Using frequency hopping, a user's time slot is not always transmitted on the same frequency. Instead, the frequency changes after every TDMA frame. This reduces the impact of destructive interference as a user's time slot is only affected when it is sent on a disturbed frequency – i.e. on average the time slot will be sent on channels which are relatively clean. The idea of synchronized frequency hopping is basically that each BTS is allocated up to 64 carriers which they can use (maximum 16 simultaneously) and the hopping is synchronized so that no two BTSs in neighbouring clusters use the same frequencies at the same time. This enables cluster size to be reduced to just four cells, and also allows the maximum number of active carriers per cell to be increased from 16 to 24.

The limited number of frequencies per cell causes another problem for the planners when areas of high population density are considered. Away from towns and cities, large

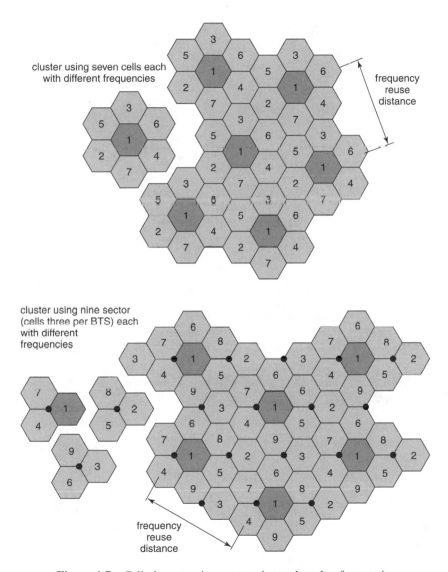

Figure 1.7   Cell clusters using seven, nine and twelve frequencies

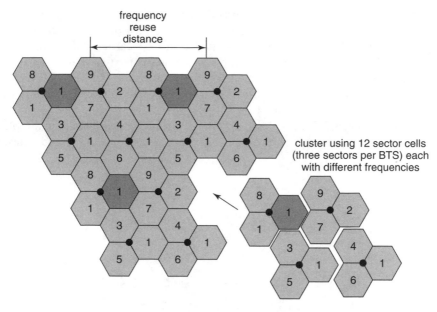

Figure 1.7 (*continued*)

cells, up to 35 km in radius, with only a few carriers can be used. As the population density increases, more carriers can be planned per BTS. If an area requires more capacity (i.e. more carriers per square km) than can be delivered by BTSs with the maximum allowed number of carriers, the operator has three main options – any or all of which can be implemented:

- smaller cells = more BTS per km^2 = more carriers per km^2,
- large overlaps = more carriers per km^2,
- sector cells = more carriers per km^2.

The first results in fewer additional problems in terms of frequency reuse. The second must be very carefully planned in terms of frequency reuse distances to avoid additional interference. The third divides each cell into sectors – e.g. $3 \times 120°$, $6 \times 60°$ (see Figure 1.8) – each of which is treated as a separate cell and also requires careful planning. The $3 \times 120°$ sector cells shown in the figure. are used to realize the nine-cell and twelve-cell clusters shown in Figure 1.7.

Once the cells and the frequencies have been planned and implemented they need to be optimized. In practice this is a never-ending task. Interference measurements need to be made and analysed, affected cells identified, and frequencies (re)allocated to improve the situation. It has long been the goal of developers to implement a system for self-optimizing networks – i.e. one in which frequencies are automatically (re)allocated based on continuous measurements. The first step towards such a feature is the so called 'smart carrier allocation' (SCA).

SCA is a software tool that enables the network operator to carry out measurements over a given period of time, allowing an interference matrix to be generated for a cell within a

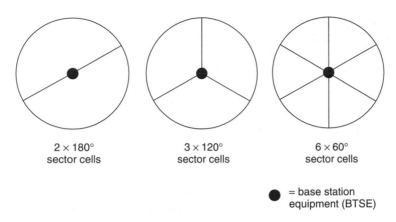

2 × 180°
sector cells

3 × 120°
sector cells

6 × 60°
sector cells

● = base station
equipment (BTSE)

Figure 1.8 Sector cells

given group of cells. The matrix allows a ranked list to be produced containing the best frequencies that can be used for the cell concerned. The cell can then be (re)configured with these frequencies. Eventually, the automation of this approach will lead to self-optimizing networks that will improve speech transmission quality and significantly reduce GPRS packet retransmission rates.

1.4 GSM – Evolutionary Concept

In 1982 the European Conference of Postal and Telecommunications administrations (CEPT) formed a team of experts, the Groupe Speciale Mobile (GSM), with the aim of developing a new standard for mobile communications in Europe. Many different technical possibilities were compared between 1984 and 1986, and in 1987 the central transmission techniques were selected. Also in 1987, 13 countries signed a Memorandum of Understanding (MoU) and agreed to start introducing GSM networks by 1991. The European Telecommunication Standards Institute (ETSI) was founded in 1988 and was responsible for converting the GSM technical recommendations into European standards. The Groupe Speciale Mobile was renamed Global System for Mobile Communications in 1989, and the following year saw the introduction of several operational GSM test systems.

The first commercial GSM phone call was made in Finland on the 1st of July 1991. That first call, from a park in Helsinki, marked the introduction of GSM900 Phase 1 in Europe. As we will see in the following sections, the development of the GSM standard continued and the specification of Phase 2 was completed in 1995. Until this point, the main focus of the development effort was on speech transmission and services related to speech calls. This was to change dramatically in Phase 2+ as we will see in Section 1.4.3.

1.4.1 Phase 1

The standardization of Phase 1 was completed in 1990 for GSM900 and 1991 for GSM1800. This initial phase introduced all the necessary infrastructure to support FR speech calls, a few basic services such as call forwarding and barring, and data transmission at rates from 0.3 to 9.6 kbit s^{-1}. Phase 1 also standardized the short message service (SMS) although initially not all network operators supported this service.

Phase 1 gave operators the opportunity to start generating revenue while development work on the GSM standard continued. This meant that they could start with a minimum of investment and expand the available capacity and service offering according to customer demand rather than waiting for the standard to be completed and implementing everything at once.

The open and incomplete nature of the standard was (and still is) the enabler for additional features (such as GPRS) which benefit the end-users and provide new sources of revenue for the network operators and service providers.

1.4.2 Phase 2

Started soon after the completion of Phase 1, the standardization of Phase 2 was completed in 1995 and brought some major enhancements to GSM. A full suite of supplementary services (such as advice of charges and multi-party calls) like those offered by ISDN networks was added, and the half rate (HR) speech codec (as described previously) was introduced. These new additions meant new mobiles had to be produced to make the features available to subscribers. At this point the decision on 'downward-compatibility'

Table 1.1 Enhancement features

Phase 2+ addition	Explanation
ASCI: Advanced Speech Call Items	Enables two kinds of group call to be made on a common downlink rather than conference call basis – i.e. originator has the UL, listeners share the DL
SOR: Support of Optimal Routing	Removes the expense of 'tromboning' – i.e. local call completion rather than via home PLMN between roaming mobiles in the same visited PLMN
UUS: User-to-User Signalling	Enables SMS messages to be sent and received during a speech call. This gives the user more flexibility and reduces queuing at the SMS center
EFR: Enhanced Full Rate	A new speech codec which offers better speech quality through added redundancy
CAMEL: Customized Applications for Mobile Network Enhanced Logic	Allows home PLMN IN services to be accessed from other PLMN's while roaming – both networks must have the CSE–CAMEL Services Environment
HSCSD: High Speed Circuit Switched Data	Implements time slot bundling and a better codec (14.4 kbit s^{-1}) but has the disadvantage of requiring a fixed connection to be set up (circuit switched)
GPRS: General Packet Radio Service	Implements time slot bundling and four new codecs and has the advantage of being packet oriented – i.e. no call setup and no connections
EDGE: Enhanced Date rates for the GSM Evolution	A new modulation scheme (8PSK) which dramatically increases the bit-rate over Um and can be combined with GPRS and HSCSD
SIM tool kit	Adds extra capabilities to the SIM card. Applications running on the tool kit can interact with the user and issue commands to the mobile to realize new services

was made – i.e. that no matter what new enhancements are introduced, the networks must continue to support mobiles from all previous phases of GSM.

1.4.3 Phase 2+

Phase 2 represented a complete update of the GSM standard, hence the need for downwards compatibility. Phase 2+ represents the start of a different approach to evolving the standard. New recommendations addressing specific topics are developed separately and added to the standard as annual releases. The GPRS specifications were added in this way along with ASCI, SOR, UUS, EFR, CAMEL, HSCSD, EDGE and the SIM tool kit. Table 1.1 briefly explains each of these features.

A common misconception is that WAP also belongs to Phase 2+. Although the wireless applications protocol (WAP) did arrive during the Phase 2+ time frame, it is not part of the GSM specifications. WAP, as explained in a later section, was designed to enable mobile Internet access, independent of the access technology (i.e. GSM, GPRS, UMTS, etc.). WAP interfaces a highly efficient access protocol with the 'resource hungry' TCP/IP protocols used on the Internet – TCP/IP uses up to 40 % of the bandwidth for flow control, which is obviously undesirable over the air interface.

1.5 The Standards

The following is a brief guide to the GSM standards and the organizations involved.

1.5.1 ETSI

ETSI was founded in 1988 shortly after the single European market was established. The European Commission responded to a growing need for an integrated communications infrastructure by recommending the establishment of an organization to set telecommunications standards for the whole of Europe. Today, ETSI plays an international role in developing a wide range of standards and other technical documentation in telecommunications, broadcasting and information technology. ETSI has 912 members in 54 countries worldwide (2002) and works closely with other standardization bodies such as CENELE (European Committee for Electro Technical Standardization) and CEPT (European Conference of Postal and Telecommunications Administrations).

ETSI receives the recommendations from workgroups (see 3GPP below) and converts them into European standards which can be found on the official web site: *www.etsi.org*.

1.5.2 The GSM Recommendations

In order to facilitate easier access, the GSM recommendations are organized into series. Information on a particular topic can be found by first referring to the relevant series. Table 1.2 can be used as a reference to locate specific recommendations.

The GPRS specifications can be found in the relevant GSM series – e.g. the GPRS service description version 7.80 can be found under 03.60 in the GSM 3 series specifications.

1.5.3 3GPP

The further development of the GSM standard, including GPRS and EDGE, has been taken over by the Third Generation Partnership Project. The 3GPP was founded in 1998

Table 1.2   GSM recommendation

Series	Description
01	The terms, definitions and abbreviations to be used throughout the GSM recommendations
02	Services
03	Network aspects
04	BTS/MS interface
05	The physical air interface
06	Speech coding
07	Terminal adaption
08	BTS/MSC interface
09	Network transitions
10	Service transitions
11	Technical conformity specifications
12	Operation and maintenance
21 to 55	UMTS

to produce globally applicable technical specifications and technical reports for a third generation mobile system. To obtain a formal status as international standards, the specifications and reports have to be converted by ETSI and other standardization bodies. 3GPP has its own web site at *http://www.3gpp.org* and its specifications and reports can be found there. Alternatively, the official ETSI versions of the 3GPP documents can be obtained from the Publications Download Area of the ETSI web site *http://pda.etsi.org/pda.*

2

The General Packet Radio Service

2.1 GPRS Objectives and Advantages

As we have seen, there were definite trends leading from speech towards data transmission and from fixed to mobile networks. Thus the introduction of a data transmission service for the GSM standard appeared to be imperative. This service would have to achieve certain objectives, both for users (U) and for network providers (NP). It must:

(1) enable access to company LAN and the Internet (U),
(2) provide reasonably high data rates (U),
(3) enable the subscriber to be reachable at all times – not only for telephone calls but also for information such as new emails or latest news (U),
(4) offer flexible access, either for many subscribers at low data rates or few subscribers at high data rates in order to optimize network usage (NP),
(5) offer low cost access to new services (U + NP).

At the time of its introduction, GPRS was the only service to achieve all of these goals. Other GSM or non-GSM technologies were failing in one or more of the points given above. Some examples are the high speed circuit switched data service offered by GSM (HSCSD), and the 3G universal mobile telephone service (UMTS). HSCSD, because of its circuit switched nature, can neither provide continuous reachability nor flexibility as in points 3 and 4 respectively. UMTS does offer continuous reachability but was not ready at the time of the introduction of GPRS and, because of high licence and investment costs, was struggling with the cost aspect – point 5.

The major advantages of GPRS arise from the fact that it uses *packet* switching:

- since devices are able to handle packet traffic, data can be exchanged directly with the Internet or company intranets;
- packets from one user can be transmitted via several air interface time slots (bundling);
- time slots are not used continuously and can be shared between users;

GPRS Networks. G. Sanders, L. Thorens, M. Reisky, O. Rulik, S. Deylitz
© 2003 John Wiley & Sons, Ltd.   ISBN: 0-470-85317-4

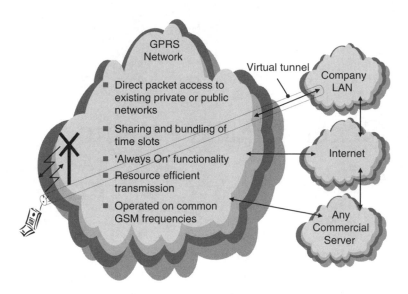

Figure 2.1 GPRS applications and strengths

- even if users are not sending or receiving, they can still remain connected/reachable (e.g. to/from their company LAN) and in doing so, they do not use up any resources;
- users are only allocated resources when they actually need them;
- GPRS is implemented within the GSM standard so no new frequencies have to be used.

Initially, the only major problem faced by GPRS was that the existing circuit switched GSM networks had to be extended to include packet switching equipment. Also, existing mobile devices were not capable of supporting packet oriented data transfer and time slot bundling so new GPRS mobiles had to be developed.

The strengths and applications of GPRS are summarized in Figure 2.1.

2.2 GPRS Architecture

The standard GSM network has had to be altered to be able to support GPRS, since it is not able to transmit data in packet switched mode. Let us first have a look at the original Phase 2 GSM network in order to be able to recognize the differences in a GPRS enabled Phase 2+ network.

2.2.1 GSM Network

A GSM network consists of many functional entities and their respective standardized interfaces. In part, a mobile network is similar to a fixed telephone network (PSTN) – the main difference to fixed networks being the almost unlimited mobility that is offered to subscribers by the use of the air interface.

A GSM network is known as a PLMN (public land mobile network) and can be divided into various parts known as 'subsystems', each of which carries out different functions. The first major division is between the interconnected network of digital tele-phone exchanges and the network of radio transceivers that enable wireless communication with the mobile. The network of exchanges is known as the 'network switching subsystem' (NSS) and is similar to a PSTN but differs in that it also includes the databases necessary for managing the mobility of the subscriber. The transceiver network (and associated net-work elements) is known as the 'base station subsystem' (BSS). The BSS together with the mobile devices and the air interface is known as the 'radio subsystem' (RSS).

The various elements described in the following sections, and their relationships to each other, are illustrated in Figure 2.2.

2.2.1.1   The Mobile Station

On the user side there is a mobile device know as the 'mobile station' (MS) which consists of the 'mobile equipment' (ME) and the 'subscriber identity module' (SIM) – as illustrated in Figure 2.3.

The ME is the device known as a cellphone or mobile telephone, and can even refer to fixed terminals such as those found in cars or remote phone booths. It consists of speaker, microphone, display, keypad and, most importantly, sender/receiver unit and antenna. The ME is able to establish both speech and data connections. For GPRS, the ME has to be equipped with packet transmission abilities.

There are three different classes of GPRS equipment:

• Class A: equipment that can handle voice calls and packet data transfers at the same time;
• Class B: equipment that can handle voice or packet data traffic (but not simultaneously) and can put a packet transfer on hold to receive a phone call;
• Class C: equipment that can handle both voice and data, but has to disconnect from one mode explicitly in order to enable the other.

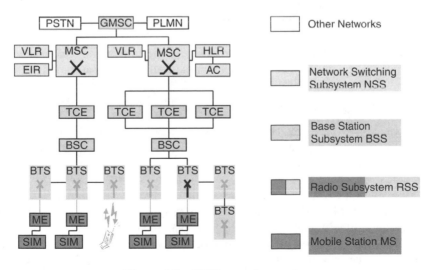

Figure 2.2   GSM network overview

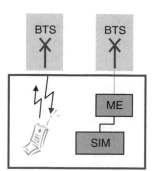

Mobile equipment (ME), device as:
1. User interface
•   speech input and output
•   data handling (received e.g. via CSD or GPRS)
•   dialling and storing numbers
•   advanced menu functions
2. Transmitter/receiver to/from BTS

Subscriber identity module (SIM), for:
1. Authentication key and algorithms
2. Ciphering algorithms
3. User identification
4. Additional telephone book

Figure 2.3   GSM/GPRS mobile stations

The SIM is provided by the network operator and usually consists of a small chip card (the SIM card), which contains important data concerning the subscriber and additional information, such as:

- International mobile subscriber identity (IMSI), which serves as a user name to identify the user towards the network. It consists of the mobile country and the network code (obviously provider dependant) and the MSIN (mobile station identity number).
- An individual subscriber authentication key, Ki, which serves as a password for the subscriber and is stored only on his SIM card and in the related database in the network (the authentication centre AC, see below).
- The ciphering key, Kc, which is calculated anew for every individual connection in order to prevent eavesdropping.
- Some mathematical procedures (algorithms) to calculate the respective keys.
- User phone book.
- A processor (SIM toolkit) for running provider specific software to enable new services such as mobile banking applications, stock market information, and more.

2.2.1.2   Base Transceiver Station

All communication (voice, fax, data or signalling) between the mobile station (MS) and the PLMN must go via a base transceiver station (BTS) – often referred to as a 'base station'. As each BTS serves a cell of limited size, a large number of them has to be deployed across the country in order to provide (ideally) seamless coverage. The speech channel capacity of a GSM network, in any particular area, is determined by the number of frequency carriers per cell and the cell density within that area. The former can vary between one and 16, with eight speech channels per frequency carrier (minus signalling channels). The latter depends highly on population density, with many more cells per unit area in more densely populated regions and fewer BTS (with relatively high output power) in rural areas (see Chapter 8, Planning and Dimensioning for more information).

Thus a BTS has to fulfil certain requirements:

- reliability in order to keep service and maintenance costs low for remote sites;
- low price as thousands of them have to be deployed;
- rugged design to withstand wind and weather on site.

The BTS has to serve as a counterpart for the mobile station (MS) on the air interface. The BTS must also interface with the PLMN via the base station controller (see next section). The main tasks of a BTS are:

- channel coding (using FR, HR or EFR),
- ciphering and deciphering (for circuit switched connections only),
- synchronization of several MS in time and frequency,
- burst formation, multiplexing and HF modulation for all active MS,
- evaluation and optimization of the UL and DL transmission quality (using own measurements and MS measurement reports).

The main roles of the BTS and its place in the BSS architecture are shown in Figure 2.4.

2.2.1.3   Base Station Controller

It would be impractical to equip each BTS with the capability to evaluate all information necessary for the handover of a particular subscriber from one cell to another. A BTS can monitor the received signal strength of the subscriber as it decreases and (theoretically) it should be possible for the BTS to estimate which cell would be most appropriate to hand over to (based on information from the measurement report transmitted by the MS). It would be impractical to equip each BTS with the capability to evaluating all information necessary for the handover of a particular subscriber from one cell to another. This would require every BTS to:

- assess the measurement report transmitted by every mobile during every call to determine whether a handover was necessary.
- have access to information on the availability of resources in each neighbouring BTS;
- be able to communicate with the neighbouring BTS to request resources for each mobile needing a handover, and to coordinate the handover itself;
- coordinate the handover with the telephone exchange so that it knows which BTS is handling each call.

There is thus a need for another network element that controls several cells (i.e. several BTS) and all the active MS in those cells. This element is the base station controller (BSC). Apart from the hand over decision mentioned above, the BSC has to carry out several other functions:

- evaluation of signalling between MS and exchanges (core network);
- performing paging in a group of cells for every mobile terminating call (MTC);
- radio resource management for each BTS;
- switching of time slots from the core network to the right BTS (and vice versa);

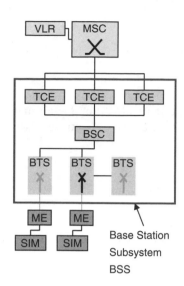

Transcoding equipment (TCE):

- Transcoding of speech data from 13 kbps⁻¹ to 64 kbps⁻¹ and vice versa.

- Rate adaptation from BSS to NSS

Base station controller (BSC):

- Simple switching tasks
- Controlling of BTS and TCE
- Signalling end point
- Operator access point

Base transceiver station (BTS):

- Transmission of compressed and ciphered data
- Modulation
- Control of transmission quality
- Power control and timing advance

Figure 2.4 Base station subsystem in detail

- providing a central operation and maintenance access point for the whole BSS;
- storage of configuration data for all elements in the BSS.

The main roles of the BSC and its place in the BSS architecture are shown in Figure 2.4.

2.2.1.4 Transcoding Equipment

There is another network element in the RSS that has not been introduced yet. As mentioned in the introduction to the GSM network, there is a conventional network switching subsystem (NSS) that consists mainly of standard fixed network components (we will consider the differences later). On the other side there is the radio subsystem (RSS), which includes the air interface between antennas and mobile devices.

Because of certain physical restrictions, the data rates for speech on the air interface are much lower than the data rate used to transport fixed network speech (the well known 64 kbit s⁻¹ from ISDN). The low rate can only be achieved by compression of the original digitized speech data. The MS digitizes speech at a rate of 104 kbit s⁻¹ (160 samples of 13 bits every 20 ms) which is compressed to 13 kbit s⁻¹ using full rate (FR) coding, 12.2 kbit s⁻¹ using enhanced FR (EFR) and 5.6 kbit s⁻¹ using half rate (HR).

The high compression ratios required for transmission via air interface time slots are a problem because the NSS uses standard digital exchanges and PCM30 lines, which make use of 64 kbit s⁻¹ time slots (it should be noted that in ISDN networks the telephones compress speech into 64 kbit s⁻¹ time slots; and for analog telephones the signal is digitized and compressed at the exchange). Because of the above, there needs to be another piece of equipment which converts the coding used for the air interface into the coding used for ISDN (and vice versa) and in doing so, converts the 13, 12.2 or 5.6 kbit s⁻¹ bit rate of air interface time slots to the 64 kbit s⁻¹ used in the NSS and other networks. This is the transcoding equipment (TCE), often referred to as the transcoding and rate adaptation unit (TRAU) as it is a data rate adaptor as well.

The question of where to position the TCE is an interesting one. The transmission of compressed data is theoretically only necessary between MS and BTS (this is where the physical restriction occurs). However, it is also advantageous to transmit compressed data through the network rather than uncompressed data, because of the line capacity which is saved. Thus the decompression does not take place in the BTS.

On the other hand, the conventional equipment in the NSS is not able to switch data compressed using non-ISDN coding (i.e. only 64 kbit s^{-1} is allowed), so the transcoding must be performed before the time slot reaches the first telephone exchange. Hence, transcoding will be done right between the BSC and the exchange – i.e. as late as possible, as early as necessary. The TCE belongs logically to the BSS (see Figure 2.4) but is usually placed close to the telephone exchange. This is done because for every single PCM30 line coming from the BSC requires four PCM30 lines from the TCE towards the network so operators can save three PCM30 lines by positioning the TCE at the exchange.

This fact in itself is not important for GPRS. What is important is the way the PCM30 lines are utilized. Each PCM30 timeslot can transport data at a rate of 64 kbit s^{-1}. Air interface time slots transport digitized speech data at a maximum rate of 13 kbit s^{-1}. This means that if the PCM30 time slots are sub-divided into 4×16 kbit s^{-1}, then four connections can be carried by one time slot. This approach combined with positioning the TCE at the exchange, is very cost effective for the network operator but causes problems when GPRS is implemented.

The problem arises because although the 16 kbit s^{-1} data rate is sufficient for all speech codecs (coder/decoder algorithms) in GSM, it is not sufficient for all codecs used in GPRS – i.e. since the GPRS data are sharing these lines at least as far as the BSC, GPRS has to be restricted to a maximum of 16 kbit s^{-1} as well. As mentioned earlier, there are new coding schemes (CS) for GPRS that are able to transmit up to 21.4 kbit s^{-1} per channel (so called CS 4) over the air interface. Obviously this can not be implemented in such a line saving environment. Even CS 3 with only 15.6 kbit s^{-1} per channel can not be used because signalling has to be transmitted on these lines as well.

As a result, early GPRS networks will only be able to use CS 1 and CS 2 (9.05 and 13.4 kbit s^{-1} per channel) until the 'bottle-neck' (illustrated in Figure 2.5) between the BTS and the BSC is removed.

2.2.1.5 Mobile Services Switching Centre

Until now we have only considered the elements in the base station subsystem (BSS). The role of the BSS is to provide access to the network switching subsystem (NSS) that enables connections to be set up between subscribers via interconnected telephone exchanges. A telephone exchange is a functional unit in the network which can:

- evaluate the dialling information (digit translation),
- find a route to the destination,
- switch the call towards the destination.

This functional unit is known as a switch or exchange, and in GSM it is known as the mobile services switching centre (MSC). Apart from the call processing tasks mentioned above, it also:

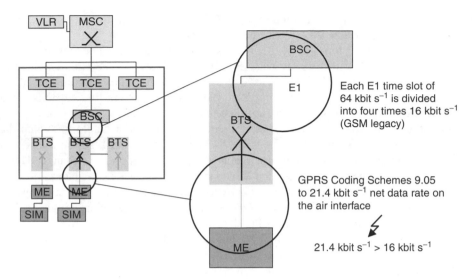

Figure 2.5   Restrictions on early GPRS networks because of GSM legacy

- creates and stores charging data records for subscriber billing,
- performs call supervision and overload protection,
- supports operation and maintenance functions,
- enables legal interception.

In case a call is made to or from another network (such as a fixed telephone network or another GSM network) there needs to be an exchange which serves as a gateway between the two networks. In this case, a special MSC will serve as a gateway MSC or GMSC.

The main roles of the MSC and its position in the NSS architecture are shown in Figure 2.6.

2.2.1.6   Location Register

So now it should in principle be possible to make a call from one subscriber to another, because we are able to transport circuit switched speech traffic via BTS, BSC, TRAU to an MSC and back to another subscriber – but there remains one very significant problem to overcome: the mobility of the subscriber.

In a fixed network, a call is possible as soon as we provide access to the network (as the RSS does for the mobile subscribers) and install an exchange (like the MSC). In mobile networks however, extra infrastructure has to be put in place to enable the correct MS to be located and contacted when a call is made to the subscriber.

So how do we locate the destination MS of a mobile terminated call (MTC) when the subscriber can move whenever and wherever he wants, and all we have is his telephone number (MSISDN)? The most common answer to this question is: 'The mobile books itself into a cell.' This statement in itself is partly correct, but it does not answer the question. The network must be able to find any MS, no matter where it is, and to do this some location information must be stored somewhere in the network for every MS.

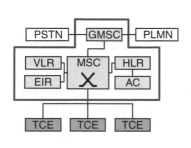

Network
switching subsystem
NSS

Mobile services switching centre MSC:
- Switching of the users
- Number translation for routing and charging
- Traffic observation
- Lawful interception
- Operator access point
- Transition to other networks (as gateway MSC)

Home location register HLR:
- Static user data
- Temporary user data like current VLR area

Authentication centre AC (associated with HLR):
- User authentication and delivery of the coding key on request of VLR

Visitor location register VLR (associated with MSC:
- Temporary user data like location area
- Requests triples from AC and stores them temporarily

Equipment identity register (optional) :
- Authentication of equipment
- Supervision of suspicious devices

Figure 2.6 Network switching subsystem in detail

A common follow up to the first statement is, 'we need a central database, which is informed (updated) whenever a user moves to another cell and can provide the destination for incoming calls'. This is again partly correct but it is important to understand the real system in order to appreciate later how GSM and GPRS can work together to manage the mobility of the subscribers.

There actually is a central database that can give information about the location of a subscriber. This database will always be interrogated when a particular subscriber is called. It is called the home location register (HLR). Apart from temporary data related to the current location of a subscriber, it also contains semi-permanent user data such as name, address, etc., for a certain MSISDN.

There are two further problems which must be addressed with regard to the storage and retrieval of location information. Firstly, a large network provider covers a whole country and may have upwards of 25 million subscribers. So if there really were only the one central database, there would be signalling messages originating throughout the whole country and concentrating at the database site whenever any of the 25 million subscribers moved to a new cell. The above scenario is unacceptable as a model for network signalling traffic as well as for database efficiency. The following presents the solution used by GSM networks.

Initially, several HLR could be introduced, each of which is responsible only for a certain fraction of all subscribers. This approach is used in GSM networks, so that each HLR has to maintain the data of about 1–4 million subscribers, depending on the preferences of the network provider. The particular HLR in which a particular subscriber is stored is coded in the MSISDN number – i.e. when a GMSC receives a telephone number it can immediately contact the HLR of the subscriber. The ID of a subscriber's HLR is given by the first two digits following the network provider code – e.g. in the number 0175 9316585 the digits '93' indicate which HLR is responsible for the subscriber.

Let us now consider how a MS is located so that the subscriber can receive an incoming call. If the network knows which cell the MS is currently in, a signalling message can be broadcast in that cell and, if the MS is switched on, it will receive the message and respond with an acknowledgment. Such signalling is known as 'paging'. The problem with this scenario is that the MS must update the location database every time it changes cell. If the operator has 25 million subscribers, the signalling load will be far too high. At the other extreme, if mobiles never perform updates, then each MS must be paged in every cell in the network whenever a call comes in. This is also an unacceptably high signalling load. The solution is to define groups of cells known as 'location areas' and each cell broadcasts the identity (LAI) of the location area it belongs to. The MS only performs an update when it moves to a new location area (determining the optimum size of a location area is explored in Chapter 3, Section 3.5.1.2). This leads us to the second question which is whether or not the HLR really needs to know whenever any of its 1 to 4 million users moves to another location area?

In reality, there is another network element which is responsible for storing the current location area of each MS. This is the visitor location register (VLR). Every MSC has a VLR which stores the LAI of every MS which is 'visiting' the region covered by the MSC. When an MS registers with the network and its LAI is stored in the VLR, the VLR identity is sent to the HLR, i.e. the HLR always knows which VLR the MS is stored in, no matter where in the world that VLR is located.

So we can now appreciate that:

- a country is divided into (several thousand) cells;
- cells are grouped into location areas;
- a number of location areas form the service area of one MSC/VLR;
- whenever a user changes a cell within a location area nothing will occur;
- when a location area border is passed the MS will notify the VLR;
- if the MS changes MSC/VLR service area, the new VLR identity is sent to the HLR.

For an incoming call, all that is known is the MSISDN of the called party, so the HLR will be asked for the location. All the HLR knows is the identity of the current VLR to which it will redirect the call set-up request. The VLR will know the location area of the called subscriber and will direct the BSC to page him in this group of cells.

The main roles of the HLR and VLR and their positions in the NSS architecture are shown in Figure 2.6.

2.2.1.7 Authentication Centre

If an MSISDN was all that was needed to make a phone call, it would be like leaving an open door to fraudulent network usage, i.e. it would simply be a case of faking the MSISDN of any known subscriber and calls would be charged to that subscriber's bill. In order to prevent such abuse, GSM implements a very thorough authenticity check of a user or, more precisely, of the SIM card in use. As with most computer systems, a

SIM card needs to log on to the GSM system with a user name and a password. To keep this system secure, the user name and password are changed after every authentication so that even if fraudsters manage to somehow intercept these two identities, they will not be able to make use of them. The procedures and identities used are discussed in detail in Chapter 3, Section 3.1.1. For the purposes of this section, it is enough to know that the authentication procedure is based on sets of keys called 'triples', which are used by both the VLR and the MS for authentication purposes. These triples are generated by a network element known as the authentication centre (AC). The triples from the AC are used to authenticate all users who attempt to make calls, receive calls, send an SMS, or even just switch their equipment on. Only emergency calls require no authentication.

The main roles of the AC and its position in the NSS architecture are shown in Figure 2.6.

2.2.1.8 Equipment Identity Register

In GSM, it is also possible, to check the status of the mobile equipment (ME) whenever it accesses the network. The word 'status' can be taken to mean legitimate, stolen or suspect. The checking is carried out by the equipment identity register (EIR) by comparing the international mobile equipment identity (IMEI), provided by the ME, against several lists. The EIR contains either a black list, where stolen equipment can be listed and eventually blocked, or a white list where legitimate equipment can be listed for unlimited use. If an ME (IMEI) does not appear on any of these lists, it can be stored on a so called 'grey' list. ME on the grey list can be restricted to certain services and all activity can be logged in detail. In case it later appears on the black list the logged information could be used to determine the location or user of the possibly stolen equipment.

Very few network providers use the EIR, as checking the IMEI of every ME every time the network is accessed generates a large volume of signalling which represents unpaid overhead for the operator. Added to which is the fact that most countries have several GSM operators, and if one operator chooses not to check for stolen equipment, then stolen ME from other operators can be used on that network simply by replacing the SIM card. Hence, many providers chose not to incur the extra signalling load required for EIR use.

The main roles of the EIR and its position in the NSS architecture are shown in Figure 2.6.

2.2.2 New Elements for GPRS

As mentioned, the general packet radio service (GPRS) is completely different from standard (circuit switched) data or speech data transmission, as the data is sent in the form of packets. Whenever a packet needs to be transmitted, the resource (bandwidth on the air interface) is allocated and the respective data packet delivered to the recipient (which for example, can be another mobile user or a server on the Internet). The network elements we have introduced so far are traffic switching, signalling nodes or both. In GSM, traffic is circuit switched with certain data rates depending on the application. Signalling is packet switched but with a much lower data rate. However, none of the nodes we have seen so far are able to transmit packet switched traffic at high data rates as required by GPRS.

2.2.2.1 GPRS Support Nodes

Clearly, a completely new set of switches is needed for GPRS, capable of routing packet switched data in a separate parallel network. There are two different main functional entities to be considered: the gateway to external data networks, and the node that serves the user.

The first is the access point for an external data network and is known as the gateway GPRS support node (GGSN). The GGSN is capable of routing packets to the current location of the mobile. Therefore it has to have access to the HLR in order to obtain the required location information for mobile terminating packet transfers.

The second is the node that serves the mobile station's needs and is known as the serving GPRS support node (SGSN). The SGSN establishes a mobility management context for an attached MS. It is also the SGSN's task to do the ciphering for packet-oriented traffic, since the base transceiver station is only responsible for ciphering circuit switched traffic. From the specifications, the two functionalities can be combined into one physical node (useful for early deployment of GPRS), separated in the same or even different networks.

Both of the GPRS support nodes (GSN) also collect charging data. Details about the volume of data (kbytes or Mbytes) transferred by the user are collected by the SGSN. Charging information detailing subscriber access to public data network (PDN) is collected by the GGSN.

2.2.2.2 Data/Voice Differentiation in the BSS

A GPRS cell phone will transmit data in a packet switched mode, but voice will be transmitted as circuit switched calls. This means that there has to be a network element that distinguishes between the two different kinds of transfer and diverts each to the correct core network, i.e. towards the MSC or towards the SGSN. This is one of the main functions of the packet control unit (PCU). Other PCU functions include:

- packet segmentation and reassembly, on the downlink and uplink respectively,
- access control,
- scheduling for all active transmissions including radio channel management,
- transmission control (checking, buffering, retransmission).

The PCU can be placed at various different positions in the network: at the BTS site, in the BSC or right before the switch, the preferred location being at the BSC.

2.2.2.3 Channel Control Unit

There is one more issue to address, namely that of how the BTS differentiates between GPRS users and non-GPRS users. There has to be some recognition of the user's needs as he/she will make normal phone calls, send and receive bursty packet traffic, share time slots with other GPRS users and will make use of GPRS coding schemes CS 1 to CS 4, depending on the quality of the air interface. These coding schemes also use their own special error correction in order to match the gross data rate on the air interface.

When implementing GPRS in an existing GSM network, every BTS requires a software upgrade so as to be capable of performing GPRS specific functions such as the use of CS 1 to CS 4, support of the GPRS radio channel measurements. This software upgrade

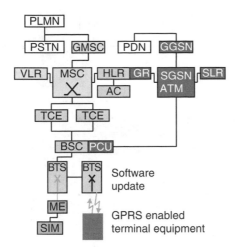

Overview of changes in the base station subsystem and core network

- Gateway GPRS support node (GGSN) for transition to the Internet

- Serving GPRS support node (SGSN) for serving users

- GPRS register (GR) as an HLR extension for data service specific information

- SGSN location register (SLR) like VLR but for data services

- Packet control unit (PCU) for the discrimination of packet and speech data

Figure 2.7   Overview of changes in the base station subsystem and core network

is specified as the channel codec unit (CCU) and is always located in the BTS. It can be necessary that the CCU takes over some of the PCU radio management functions if the latter is not located in the BTS itself.

The overview of a Phase 2+ network described above is illustrated in Figure 2.7. and the functionality of all the new elements will be covered in detail in Chapter 6.

2.2.3 Interfaces of the Network

Now that we have introduced all the new network elements, we can have a look at the whole picture. Obviously, the new network elements require new interfaces to be implemented. The new interfaces interconnect both new and existing network elements. Some of these interfaces are essential for the introduction of GPRS while others are specified but will be implemented in later releases of the GPRS standard as new functionality is introduced, e.g. early releases do not implement the Gc interface as they do not support mobile terminating packet transfer. All the interfaces in GPRS are labelled 'G' followed by a letter to indicate a particular interface.

The main interfaces for GPRS data transmission as illustrated in Figure 2.8 are:

- Gb, the interface between the BSS (PCU) and the SGSN,
- Gn, between all SGSN and also to the GGSN in a mobile network,
- Gi, the interface from the GGSN to a public data network,
- Gp, an interface to the GGSN of another mobile network.

There are also several signalling interfaces between GPRS nodes and also towards standard GSM network elements.

- Gr, the interface from SGSN and Gc, from the GGSN to the HLR for the exchange of user subscription, service and location data (in case of GPRS roaming, the HLR can be in a different network or even country);

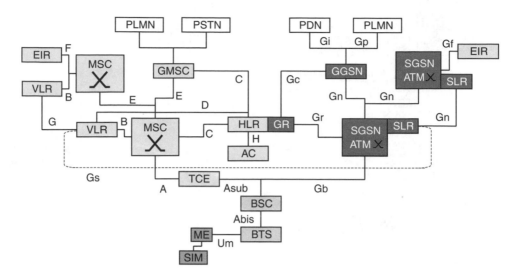

Figure 2.8   Overview of GSM and GPRS network interfaces

- Gd, an interface towards an MSC or GMSC with a connection to the SMS centre;
- Gs, an interface to an MSC to enable common mobility management (explained in detail in Section 2.4 Logical Functions);
- Ga, connection between SGSN, GGSN and the charging gateway function for the collection of billing data;
- Gf, the optional interface to an EIR for an extra equipment check in GPRS.

The interfaces, their functionality and the protocols used will be covered in much more detail in Chapter 3 Interfaces and Protocols.

2.3   Characteristics of a GPRS Connection

Let us have a look at the characteristics of GPRS data transmissions, in particular at how data is transmitted in GPRS, how higher data rates than with classical circuit switched data (CSD) are achieved, and especially how we can still keep it flexible enough to provide acceptable quality of service (QoS) for many users. This section aims to give an overview of the requirements and specified mechanisms in order that the reader can better understand the physical details that follow in Chapter 3 Interfaces and Protocols.

One of the major advantages of using packet switched transmission is the flexibility it offers to both subscribers and operators. When a single user needs to download data, he or she can make use of the entire available bandwidth. If there are other users, they can share the bandwidth according to their individual needs or priority level. In a circuit switched connection, a resource (the line) is allocated to just one user for the duration of a call or data transfer, regardless of whether he actually makes use of it or not. This usually means that there is a call set-up procedure before the call (a procedure to allocate the resources so that they are actually reserved for the user for the duration of the call). When the connection has been established (set-up), there is nothing further for the network elements to do but maintain the call, perform handovers if necessary, and collect charging data. This is completely different from a packet switched connection in which the resources are shared and there is no guarantee that there will not be delays in the transmission (making such guarantees forms part of implementing QoS). For example, one user could be using the resources when another one tries to access it and unless the second user has a higher priority his or her transmission will be delayed. Another difference is the addressing, i.e. defining the origin and destination. In the absence of a call set-up procedure, the network must have a method of determining the recipient of the current data packet. Hence, each packet has to carry additional addressing information. The subscribers must be charged for the use of network resources. Since each time slot can be used intermittently and shared between several users, the billing cannot be based on time. Thus a new billing model has had to be developed based on the volume of data transmitted by each subscriber.

2.3.1 Coding Schemes and Management of Radio Resources

Now let us have a look at how the packet transmission is realized in GPRS. As we have already seen, GSM uses eight time slots per carrier frequency. If an MS transmits data with the classical GSM circuit switched data (CSD), then one of these time slots is allocated to the MS for the whole duration of the data connection. Each time slot offers a gross data rate of 22.8 $kbit\,s^{-1}$ but not all of this bandwidth is available to transfer user data. To ensure that the data is transmitted reliably, a copy of the data is made and sent separately across the air interface, i.e. each time slot contains both original data and a back-up copy of data from other time slots. As a result of this built-in redundancy, the net rate of a CSD connection is 9.6 $kbit\,s^{-1}$ (or less if the quality of the air interface is poor).

2.3.1.1 HSCSD

A higher data rate could be achieved by increasing the net data rate, by bundling several time slots, or both. A higher data rate means using a new coding scheme that will necessarily send more original data and as a result will have less redundancy. Theoretically, 'bundling' means that up to eight time slots can be made available to one user as this is what is available on one frequency carrier.

The above measures have been implemented in GSM and the result is known as high speed circuit switched data (HSCSD). Using HSCSD enables faster data transfer (14.4 $kbit\,s^{-1}$) but this is still just a faster type of circuit switched connection with all the associated advantages and disadvantages. In principle, HSCSD could deliver 8×14.4 $kbit\,s^{-1}$ (115.2 $kbit\,s^{-1}$), but since it is circuit switched, it will be transmitted via the classical nodes (BTS $\rightarrow$ BSC $\rightarrow$ TCE $\rightarrow$ MSC $\rightarrow$ provider) and therefore can not exceed the MSC data rate of 64 $kbit\,s^{-1}$ per connection. As a result HSCSD is practically limited to 57.6 $kbit\,s^{-1}$, i.e. four time slots using 14.4 $kbit\,s^{-1}$ (Figure 2.9).

2.3.1.2 Coding Schemes in GPRS

To some extent, GPRS makes use of a similar approach to HSCSD in order to achieve high data rates, i.e. GPRS also offers channel bundling of up to eight time slots and increased net rates per time slot by sacrificing redundancy. Despite a superficial similarity, however, GPRS is fundamentally different from HSCSD as can be seen in the following text, which considers the differences in the GPRS channel coding.

CSD

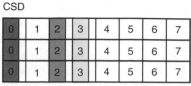

Example: three users, getting a single time slot each in every TDMA frame (0, 2 or 3). Net rate 9.6 kbit s^{-1}

HSCSD

Example: Two users, each bundling four time slots in every TDMA frame. One gets 0,1,2 and 4, the other 3,5,6 and 7.4 × 14.4 kbit s^{-1} = 57.6 kbit s^{-1}

Advantages of HSCSD:

- No changes of the GSM network necessary
- Real time ability

Disadvantages of HSCSD:

- Not resource efficient as each user occupies all his channels throughout the data connection
- Limitation to 64 kbit s^{-1} (max. four time slots) because of switching subsystem restrictions

Figure 2.9   HSCSD and classical GSM CSD time slot functionality

The design of GPRS takes into consideration the variability of the air interface and the effect this has on transmission quality. When the air interface exhibits good transmission characteristics, e.g. a very low retransmission rate, little or no redundancy is required. When, however, the rate of retransmissions increases, coding schemes with more redundancy must be utilized. As a result, GPRS was designed with four new coding schemes: CS 1 to CS 4. Coding scheme 1 has a net rate of 9.05 kbit s^{-1}. Obviously this is even less than the CSD rate of 9.6 kbit s^{-1}, but CS 1 is extraordinarily stable, even under the worst air interface conditions, because of its high redundancy convolutional coding. Coding schemes 2 and 3 offer 13.4 and 15.6 kbit s^{-1} respectively. For the actual coding they use the same algorithm as CS 1, but with the high redundancy of CS 1 they would exceed the gross data rate of 22.8 kbit s^{-1}, which is impossible. So following the convolutional coding, a process called 'puncturing' is carried out. This means that certain bits of the copy (redundancy) are deleted in a prescribed manner in order to achieve the desired data rate. As a result, CS 2 offers less reliability under bad air interface conditions (that is when the copy might be needed more often) than does CS 1. Obviously CS 3 is even more dependent on good transmission conditions.

CS 4 offers a net rate of 21.4 kbit s^{-1}, which means that this coding scheme carries no redundancy at all. Consequently, CS 4 should only be used when the air interface offers close to perfect conditions, since every transmission error results in a retransmission, which reduces the effective data rate. In practice, CS 4 will only be used when the MS is close to the base station and is either stationary or moving very slowly.

It is worth noting here that all time slots contain a checksum so that transmission errors can be detected by the receiver. This checksum is the reason that CS 4 can not use the full 22.8 kbit s^{-1}. The GPRS coding schemes and the time-slot sharing described in the next section, are illustrated in Figure 2.10.

2.3.1.3   Radio Resource Management

Because of the packet switching mode, there is no connection set-up in GPRS as there is in CSD and HSCSD data calls, and all speech calls. Furthermore, it is not known simply

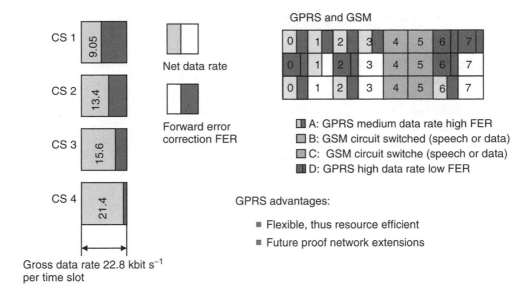

Figure 2.10 GPRS time-slot sharing and coding schemes

from the frequency carrier and the time slot number to which subscriber the transmitted data belongs, i.e. in GSM a user channel is defined by a time slot number and frequency pair (UL and DL) and this is not the case in GPRS. The assignment of carrier and time-slot number(s) to a certain user has to be managed dynamically in GPRS and this applies to both UL and DL, i.e. the allocation of UL and DL resources is managed separately, which means that while one subscriber makes use of a particular DL time slot, another can be using the corresponding UL time slot. This is done by the radio resource management functions, which will be described in more detail in Chapter 3 and is mentioned here only to introduce the concept and highlight the difference to the fixed bundling method of HSCSD.

In HSCSD, the number of time slots to be made available for the subscriber is decided at the beginning of a data call and is then allocated for the entire duration of the call. In GPRS however, following an initial 'logging on' procedure known as a 'GPRS attach', there will be no channel assignment whatsoever until the user is ready to send/receive some data. Only when there is actually a transmission request to or from the attached subscriber will it be decided how many time slots he or she will receive. If there are no other GPRS users, he or she might actually be allocated all eight time slots for the duration of the data transmission. However, if there are many users, he or she might be allocated only one or two time slots. In order to enable maximum flexibility, this decision can be revised after every four TDMA frames, i.e. one user might have several time slots continuously *or* one time slot has several users who must each wait for permission (from the PCU) to use it. GPRS makes use of a mechanism called media access control (MAC), which is described in detail in Chapter 3. It is important to note here that the MAC protocol uses 3 bits to allocate UL resources and 5 bits to allocated DL resources, which means that there can be a maximum of eight subscribers sharing an uplink time slot and 32 subscribers sharing a downlink time slot (the number actually implemented will depend on the network operator). This is what makes GPRS much more flexible than HSCSD: up to 64 subscribers can share or bundle the eight time slots of one DL carrier frequency depending on their individual needs. Theoretically they never actually have to detach themselves from the service as they do not block any resources as long as there is no transmission request. This is the famous 'always on' functionality more accurately known as a 'logical link'. The logical link is analogous to the relationship between your local post office and your home. Once the post office knows your address, a link exists between them and you which only exists as a logical concept but is utilized physically when somebody sends you something. In GPRS, the user is attached to the network but resources are only allocated when required (see Figure 2.11). This forms the basis of the GPRS 'pay for use' or 'volume billing' approach.

2.3.2 GPRS Subscriber Profile

The GPRS subscriber profile contains information about the content of a subscriber's contract. When quality of service is implemented, the operator will have to 'package' the QoS parameters in such a way that the subscriber has a simple choice to make rather than be confronted with complex parameters, i.e. the customer cannot be expected to choose a mean throughput class, reliability class or a delay class, etc. The customer must be offered levels of service, e.g. bronze, silver and gold, each of which has a certain QoS associated with it. The parameters underlying each service class will be stored in the subscriber profile along with other aspects such as IP address, screening profile, etc. Usually a service request from the user will be validated against the user's subscription profile. The profile is stored in the HLR extension known as the GPRS Register (GR). It is also possible to renegotiate a request, if the network can not fulfil the users demands.

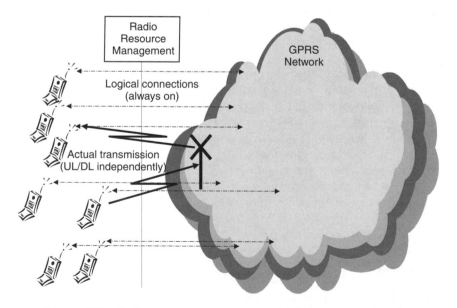

Figure 2.11   Radio resource management and 'always on' functionality

2.3.2.1 Transmission Services

Another characteristic of GPRS is that it actually offers more than just a single data service. A lot of different data services and different fixed data networks are already in existence, and GPRS was designed to give access to these networks. So just as GSM had to be adapted to support interconnection to established voice call services in digital ISDN or analog POTS networks, GPRS needs to be able to inter-work with existing networks.

In GPRS, two network services are supported (Figure 2.12):

- Point-to-point connectionless network service PTP-CLNS
- Point-to-point connection orientated network service PTP-CONS

The term 'Point to Point' (PTP) is used because data packets are always transferred from a source to a destination, only one of which has to be a GPRS subscriber, i.e. data can be transferred between users or between a user and a server (e.g. on the WWW, an intranet, etc.). In the PTP-CLNS, each data packet is routed independently of all others through the network. Therefore, no connection has to be established beforehand and each packet can find its own way through the network. This is very useful for bursty traffic, since it avoids the congestion often associated with single dedicated paths. This is a so called 'datagram' service, such as that used in the Internet or more generally in TCP/IP networks. The support of the Internet protocol IP of the TCP/IP suite was the main intention of this service. The PTP-CONS is obviously connection oriented, which means that a dedicated path has to be established between two users before they start the actual transmission. The connection may last only a few seconds or even hours. This should not be confused with a circuit switched connection (as for a voice call). PTP-CONS is just like PTP-CLNS used for bursty traffic so the relationship between the two subscribers is only a logical connection and only when they actually have to transmit data,

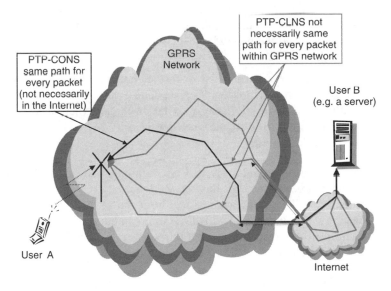

Figure 2.12 GPRS transmission services

it becomes a physical connection on the dedicated line. The circuit switched connection, in turn, is a physical link from the beginning to the end. The connection orientation can be useful for interactive applications where the delay for a reaction to some action of the user has to fulfil certain requirements. These can be planned more easily on a dedicated path (the connection orientation) than with a datagram service. It is also used when an action of the user triggers a transaction with certain requirements for reliability, safety and system response.

GPRS also supports broadcast functionality to several subscribers, which is known as 'point to multi-point'. This system takes a packet from the source and sends copies of the data in the packet to different destinations, i.e. each copy has a different address in the destination address field of the IP header but the payload is the same.

2.3.2.2 Quality of Service (QoS)

As we are already aware, GPRS is designed to allocate the scarce (radio) resources dynamically in order to be able to respond efficiently and effectively to local traffic conditions and subscriber demands. The more users there are transmitting data at any one time, the less data that can be transmitted by each one individually. The more users sharing a time slot, the longer the (potential) transmission time for each user.

The active user (of the subscribers sharing a time slot) can be changed frequently by the PCU (actually every four TDMA frames – for an explanation, see Chapter 3) so that each MS transmits in turn, i.e. many users sharing a time slot are not blocked by one user sending a large file. Even though the MS may request a certain bandwidth, the network will allocate resources on the basis of providing the best QoS to all current users. As a result, users cannot expect to get a consistently small delay or fixed data rate as with circuit switched connections. They also have to cope with the higher probability of data loss. However, this is the price that must be paid to provide maximum flexibility for a much larger number of subscribers. There are several quality of service classes that provide predictability of the data transmission for the certain applications. The class will be 'chosen' by the subscriber for the whole contractual period and has to be negotiated by every new application and the network.

Precedence Class (priority)
The precedence class (Figure 2.13) is used to further enhance the flexibility of GPRS. Packets can be marked with their respective precedence class, to indicate whether they need to be transmitted with high priority or may be discarded when necessary (in order to relieve network congestion for example), i.e. low priority.

There are three possible values:

- High precedence: Service commitments will be maintained ahead of all others, which means a premium can be charged for this service level.
- Normal precedence: Will fulfil the average requirements of the average subscriber.
- Low precedence: Service commitments will be maintained after all higher precedence levels have been fulfilled, which means that this service level will be cheaper but will carry the risk that packets may be deleted on route if necessary.

Data packets in GPRS networks are stored (buffered) for a very short time in each network node along the route between source and destination. These nodes, such as the

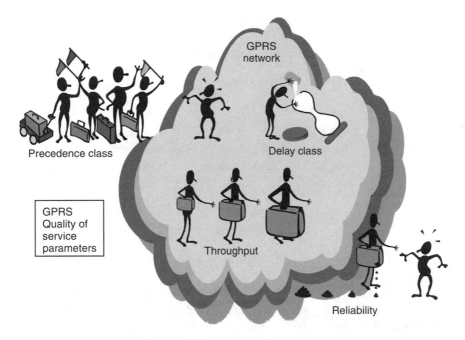

Figure 2.13   Quality of service classes

PCU, SGSN and GGSN, must continuously make decisions based on the precedence class of the subscriber, as to whether packets should be forwarded immediately, buffered until resources are available or deleted. In the case of the PCU, a subscriber's precedence class can be used to give access to shared resources on the air interface, i.e. if all subscribers sharing a time slot have the same precedence class, then statistical multiplexing can be used, whereas if one user has a higher priority, then the other users must wait.

Reliability Class
Low precedence indicates that data may be discarded in case of network congestion, but high precedence does not guarantee flawless transmission. Different applications rely differently on the reliability of the transmission. For example, the TCP in TCP/IP controls the data flow very thoroughly and initiates a retransmission in case of a packet loss. Thus, applications on top of TCP/IP, such as http or ftp, can count on perfect transmission regardless of whether packet loss occurs or not. On the other hand, an application based on UDP relies heavily on the correct transmission and should (ideally) be operated via reliable connections only. It is not easy to define different grades of reliability as common sense dictates that a transfer or connection is either reliable or not, but it is possible to define some reliability parameters and with a statistical approach over many transmissions a relative reliability can be defined, i.e. the statistical probability that a so called service data unit (SDU) will not be delivered 'problem free'. Typical examples of reliability problems are:

- The possibility, that an SDU gets lost, e.g. discarded due to a transmission error. (lost SDU probability);

- The possibility that an SDU will be delivered twice
  (duplicate SDU probability);
- The possibility that a set of consequent SDU arrive in the wrong order;
  (out-of-sequence probability);
- The possibility that a user receives a corrupted SDU without the error being detected.
  (corrupted SDU probability).

Again, GPRS offers a flexible solution to different needs by introducing three different
reliability classes, which are certain combinations of error probabilities based on the four
points listed above. The various error probabilities of the different classes are compared
in Figure 2.14.

Reliability Class 1 means that, statistically, each of the four SDU error probabilities is
   below one faulty SDU out of every billion transmitted SDUs. Thus the probability
   equates to 10^{-9}.
Reliability Class 2 sets a limits from 1 in 10 000 for lost SDUs up to 1 in 1 000 000 for
   corrupt SDUs.
Reliability Class 3 allows the Corrupted SDU probability and the Lost SDU probability
   to go up to one faulty SDU out of 100 transmitted SDU, whereas the probability of
   duplicate or out-of-sequence SDUs is the same as Class 2: 1 in 10 000.

It is up to the application to demand the required reliability class and the above values
have been selected to enable applications with different error sensitivity – e.g. streaming
video and telemetry data – to request the appropriate service class. For example, Class 1
would be necessary for applications that are error sensitive but which do not offer any
error correction themselves. Applications with good error tolerance capabilities or limited
error correction functionality (in case they are error sensitive) could cope with Class 2.

Reliability class	Lost SDU probability (a)	Duplicate SDU probability	Out-of-sequence SDU probability	Corrupt SDU probability (b)	Example of application characteristics
1	10^{-9}	10^{-9}	10^{-9}	10^{-9}	Error sensitive, no error correction capability, limited error tolerance capability.
2	10^{-4}	10^{-5}	10^{-5}	10^{-6}	Error sensitive, limited error correction capability, good error tolerance capability.
3	10^{-2}	10^{-5}	10^{-5}	10^{-2}	Not error sensitive, error correction capability and/or very good error tolerance capability.

Figure 2.14   Reliability classes in GPRS

Class 3 would only be suitable for applications that are either completely insensitive to errors or which offer a sufficient level of error correction themselves.

The network may or may not be able to grant the requested reliability class. In the case of high system load or if the requested class has not been implemented (e.g. early GPRS networks operate on a 'best effort' basis) then the request has to be renegotiated.

Delay Class

As mentioned previously, there will always be some delay in transporting the data through the network (even short storage in switching devices), and as the packets from one application can be routed via different paths, the delay will not be constant (this is known as 'jitter'). Again there are some applications that are delay sensitive, e.g. real-time applications, or interactive applications (where delays in system responses to user input cause frustration and customer dissatisfaction), and some that are not, e.g. short messages (SMS), as the user will not usually be aware of delays of up to several minutes or even hours.

In GPRS an application can select one of four different delay classes that are specified within the GPRS network only. The values actually specified are the mean delay and the time when 95 % of the data are most likely to arrive. So 5 % of all data are not explicitly specified, only implicitly, because overall the data have to match the mean delay parameter. The GPRS specifications refer to two different packet sizes – these being 128 octets and 1024 octets respectively. The delay is usually measured from user to user, or user to GGSN (i.e. not from hop to hop but for the whole transmission path) but can be guaranteed only within the GPRS system, since delays in external networks such as the Internet are unpredictable. So when external networks are accessed, the GPRS delay is valid from the GPRS gateway support node (GGSN) via the whole network and the radio interface to the GPRS mobiles.

The delay Classes 1 to 3 offer certain guaranteed delay values (see Figure 2.15). Class 4 does not specify any values, it just offers best effort transmission and thus is heavily dependent on the network load. Class 4 is the only class that has been implemented right from the start by all network operators.

Delay class	Delay (maximum values)			
	SDU size: 128 octets		SDU size: 1024 octets	
	Mean transfer delay (sec)	95 percentile delay (sec)	Mean transfer delay (sec)	95 percentile delay (sec)
1 (predictive)	<0.5	<1.5	<2	<7
2 (predictive)	<5	<25	<15	<75
3 (predictive)	<50	<250	<75	<375
4 (best effort)	Unspecified			

Figure 2.15   Delay classes in GPRS

Throughput

The throughput is divided into two different parameters, maximum bit rate and mean bit rate.

The theoretical maximum bit rate specified in GPRS is 171.2 kbit s^{-1}, i.e. one subscriber using coding scheme 4 (21.4 kbit s^{-1}) on all eight time slots on one carrier. Firstly, the transmission of eight time slots is not necessarily supported by all GPRS mobiles (see Chapter 7 for details). At the time of writing, most mobiles on the market offer four time slots on the downlink and one or two on the uplink. Secondly, CS 3 and 4 are not supported by all network providers. At the time of writing, CS 2 with a maximum transfer rate of 13.4 kbit s^{-1} is the limit. Solutions are being developed to overcome the bottleneck on the Abis interface and enable the use of the higher data rates of CS 3 and CS 4, with minimal changes to the access lines.

With or without the maximum rates offered by eight time slots and CS 4, the fact remains that network load will change dynamically according to subscriber demand and therefore it would not be in the best interests of the subscribers or the operators to provide each user with a fixed throughput guaranteed for the duration of the GPRS session. One of the main goals of GPRS has always been flexibility, and this also applies to the throughput made available to each subscriber at any particular point in time. Accordingly, the maximum throughput has to be negotiated (between the MS and the network) when the mobile attaches to the network, i.e. the mobile requests certain resources and the network decides what resources will actually be made available. As the negotiated throughput could be significantly lower than the level desired by the user (or the capabilities of the mobile), the network should also be able to alter this value at any time if a change in network load occurs. The maximum bit rate is not a very useful value for a packet switched service, as it is not actually used for long duration transmissions and does not describe the bursty nature of most packet switched applications. For this reason, a mean bit rate is also negotiated. This is a long term level of throughput that the user can rely on. It includes both short term transmissions at maximum throughput and transmission pauses when no data are pending. The mean bit rate can also be renegotiated during a session, in case the conditions in the network and especially the air interface have changed.

2.4 Logical Functions

Both GSM and GPRS have the same goal: to enable the mobile subscriber to establish communication links from a mobile device to a second party. As this common goal is enabled by the same mobile device via the same radio subsystem (RSS), we can expect that they will also have many similarities in terms of the logical functions necessary for operation, such as network access control, mobility management, radio resource allocation and collection of billing data. This is true to some extent, but we must also expect that the requirements and procedures of some of these functions can differ significantly due to the different nature of the communication involved, i.e. circuit switched versus packet oriented. Added to this, GPRS has many functions not found in circuit switched GSM. This section explores the logical functions of GPRS so that the reader can appreciate the common ground, the similarities and the differences as compared with circuit switched GSM.

2.4.1   Network Access Control

Network access can simply be described as the way in which a user becomes connected (or 'attached') to a network in order to use the services and network facilities. The reality is much more complicated than this simple statement implies, since users not only have to be registered but also be checked for their authenticity, their subscriber profiles, whether there are enough network resource to handle another user and so forth.

The term 'GPRS attach' summarizes all the necessary steps to get access to the network and can be thought of as 'logging on' to the network. The opposite expression 'GPRS detach' describes the procedures for 'logging off'. The following paragraphs will explain these functionalities in detail (Figure 2.16).

2.4.1.1   Registration

A mobile has to perform a registration operation in order to make itself known to the network. After successful registration, the network should, for example, know the current routing area of the user, the user's permanent IP address if the user has one (IPv6) and his access point names (APN), be they for the Internet, a company LAN or both. In early versions of GPRS operating under IPv4, the IP address is dynamically allocated during the PDP context activation (see Section 2.4.2.1), but later releases operating with IPv6 will store the permanent IP address of the MS in the GPRS register (GR) as part of the subscriber profile. The outcome of a successful registration should also be a valid service profile (QoS, screening profile) in the SGSN, which could be the static subscriber profile taken from the HLR or, in case of renegotiation, a dynamic one as discussed in the previous section. So as a result of a registration, the mobile station should be ready for activating a PDP context which will determine the transmission of GPRS data between the MS and a data network (access point).

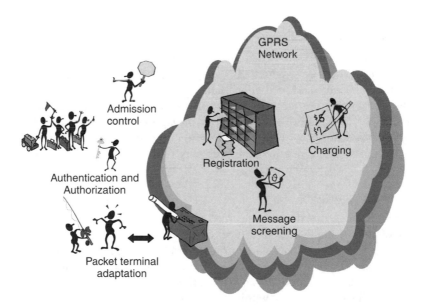

Figure 2.16   Overview of network access control functions

2.4.1.2   Authentication and Authorization

Authentication refers to the necessity for checking the identity of the MS before it is allowed to make use of network resources. Initially this procedure was intended to protect the subscriber from fraudsters who would make illegal use of the network by stealing and using their identities. With the rise of mobile services and m-commerce, the authentication procedure is seen as even more important as the SIM card is being used more and more as a credit card, e.g. it is already possible to purchase goods and pay taxi fares with your mobile telephone. The procedure involves an exchange of electronic keys between the VLR and the MS, and will be covered in detail in Chapter 4, GPRS Procedures. During the procedure an electronic signature (actually a bit sequence) is sent to the VLR from the AC. The VLR then sends another key to the MS with which the MS should be able to generate the same electronic signature as that supplied by the AC. If the two signatures (bit sequences) are the same, the authentication is successful and the MS is provided with access to the network, otherwise access is denied.

Authorization refers to the fact that following a successful authentication, the VLR is provided with a copy of the subscriber data which specifies what types of calls and services the subscriber is allowed to make, i.e. has paid for. So when a user attempts to make a phone call or use a service, the VLR checks the subscriber profile to see whether or not the user is allowed to do so.

Thus it can be said that authentication gives a user access to the network and authorization enables this user to make use of the network. A good example of this is SMS in Europe. Although many operators in different countries have not yet made roaming agreements with each other, it is possible in most countries to send and receive SMS, i.e. the roaming MS are authenticated and therefore have access to the networks but the subscriber profiles forbid them from making telephone calls as there is no roaming agreement, the only agreement is for SMS.

2.4.1.3   Admission Control Function

The admission control function is the one that calculates the network resources needed to fulfil a certain subscriber's request. It then has to check whether these resources are actually available (or not) and then allocate the available resources to the particular subscriber. The checks that the admission control function has to perform not only cover quality of service (QoS) aspects such as mean throughput, reliability class and priority, but also have to be associated with the radio resource management in order to allow for the frequency and time slot requirements for every GPRS subscriber in the cell.

2.4.1.4   Message Screening

The message screening function is concerned with filtering unauthorized and unsolicited messages. It should be implemented on a per packet basis as in a firewall, and is really just a basic filtering function for packets from undesirable sources or with undesirable content. This does not serve as the one and only security measure in a GPRS network. It can not find malicious code on an application level or prevent different kinds of attack from the Internet or from within the mobile network itself, say via a compromised GSN.

The message screening function is usually implemented on a network basis and might later be upgraded for an individually customizable screening function for every single user.

2.4.1.5 Packet Terminal Adaptation

Very often the mobile will be connected to another device on which an application is running. The mobile must therefore be able to adapt to the variety of interfaces which can be used to exchange data between itself and the external device. The packet terminal adaptation function fulfils this role, i.e. packets received via the air interface must be stored and adapted before being sent to the application running on the external device.

2.4.1.6 Charging Data Collection

There are two different functions necessary for collecting charging data in GPRS: use of the GPRS network belonging to a subscriber's HPLMN, and access to external networks.

The billing for use of the GPRS network to which a user subscribes can be based on various models, the most popular of which seems to be a subscriber's traffic volume (in kbytes or Mbytes). The charging data for traffic within a GPRS network, i.e. user to user or user to gateway, is collected by the SGSN to which a subscriber is attached.

The billing for use of external networks such as the Internet can also have different basis. For example, billing for the use of the Internet is currently based only on access but there is increasing interest in also charging for content. The charging data for access to external networks is collected by the GGSN.

As part of the 'logging on' process (see 'PDP context activation' in Section 2.4.2.1) the GGSN sends a so called 'charging ID' (CID) to the SGSN. During the billing process batches of charging data records (CDR) are regularly sent from each network node to a central billing centre. The CID is used to merge the records from different network nodes which apply to the same subscriber.

2.4.2 Packet Routing and Transfer Functions

The transfer of voice data in GSM is relatively simple to follow as it is circuit switched. There is one-to-one mapping between time slots from the RSS and time slots in the NSS, and these time slots always follow the same path from end to end. The transfer of data packets in GPRS however is not so simple to follow as the packets are very large in comparison with the time slots on the air interface, and packets moving between the same source and destination can be sent via different paths. Added to this is the problem of the different interfaces with different protocol structures over which the packets must be sent. The terminology used to describe the way packets are handled by the protocols and the way the packets are transferred must be understood in order to comprehend fully the process involved. For example, in GSM we speak of 'connections' while in GPRS we speak of 'routes'. The former is a fixed path followed by every time slot. The latter is the path followed by an individual packet which can be the same as or different from, the path followed by the next or previous packet. The logical functions on the transmission path are illustrated in Figure 2.17.

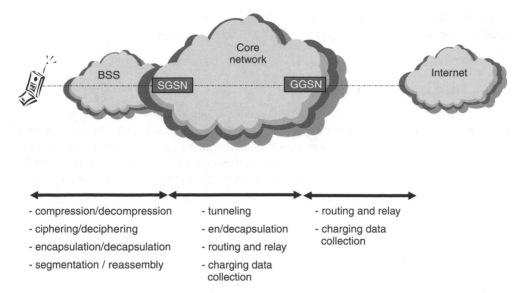

Figure 2.17 Logical functions on the transmission path

2.4.2.1 The Routing and Relay Functions

These terms refer to what happens when a packet travels across a packet data network. Described simply, the packet arrives at a node, the routing function decides which network node the packet should be sent to next in order to reach the destination eventually, the packet is sent to the next node and the above process is repeated. The relay function is the means by which a node forwards the packets received from one node to the next node which is determined by the routing function. These functions are not unique to GPRS – every packet data network must perform these functions.

Routing is often compared with sending packages through the post. In some ways this is a good analogy in that the postal system also makes use of addresses that enable the package to be transported across various infrastructures to the destination. This analogy does not, however, reflect the complexity of the routing decisions which must be made by the routing function in data networks. The movement between two nodes is called a 'hop' and routing decisions must be made on a 'hop-by-hop' basis according to a set of rules. The rules can dictate that decisions are based on the current load on each hop, the available bit rate on a hop, or other such parameters (see Chapter 8 for more examples). Data transmission between GSNs may also take place via data networks that implement their own internal routing functions, e.g. IP over Ethernet, frame relay or ATM networks (the Gn interface used between GSNs, can for example be realized as IP over ATM).

2.4.2.2 Address Translation and Mapping Function

The previous Section (2.4.2.1) mentioned that packets may be transported across several different networks in order to reach the destination. If the networks concerned do not use the same kind of addressing, then a translation function is required, i.e. the addressing information in the packet header will be used to generate a new header with an address which can be used by the second network. It should be noted that address translation can be used for routing packets within and between GPRS networks, and also between GPRS networks and external data networks.

If, however, the networks concerned do use the same kind of addressing then address mapping rather than translation is used. This means a one-to-one relationship will be established between a network address in one network and another network address in a second network, i.e. an entry in a so called 'look up' table will be made which determines where packets with a certain address will be sent.

2.4.2.3   Encapsulation

This function is one of the most important aspects of the chapter on interfaces and protocols (Chapter 3). When a packet is sent across an interface, some extra information must be attached to the packet. This extra information has various forms and various uses and can be likened to a letter being sent through the post. The letter itself must be put in an envelope with the destination address printed on it. At the local post office the letter is put in a sack which is transported to the central post office. At the central office, the contents of the sack are sorted into other sacks according to destination (city or country) and the sacks for each destination are put into containers for transportation. The transportation may involve several steps before a postal employee, who has a bag containing letters for delivery in the same street or area, finally delivers the letter. The letter, envelope and address are like an IP packet, i.e. information with an address. Putting an envelope in a sack, or a sack in a container, or a container in a boat/plane/train/truck is encapsulation – i.e. the address defines the ultimate destination but the letter is not itself in a suitable form to be individually transported through the postal system. The sacks enable easy transport by the local postal service and the containers enable easy transport to a certain destination by the national postal service. It is important to note that encapsulation can have many layers, i.e. {[(envelope) sack] container}, and that each layer is added for a specific purpose and is removed when that purpose has been achieved. In the case of data packets, additional information is added in the form of headers and check bits which 'encapsulate' the original packet. The removal of this information when it has served its purpose is called 'decapsulation'.

The protocols described in Chapter 3 will clearly demonstrate the importance of en/decapsulation – the Gn Interface between the SGSN and the GGSN is a good example of multiple layer encapsulation of an IP packet.

2.4.2.4   Segmentation and Reassembly

Depending on the physical transport media, i.e. the technology used to realize layer 1, it can be that the datagram is too large to be sent as a single transmission. In that case the datagram must be divided into several smaller segments, each of which receives a header to enable the original datagram to be reassembled on the receiving side. For example, the protocols that prepare an IP packet for transmission over the air interface make use of segmentation, e.g. the IP packets are segmented into smaller frames (called LLC frames) which, in turn, are segmented into radio blocks, each of which is then divided into four time slots (see Chapter 3).

2.4.2.5 Compression

In order to optimize the use of the air interface, i.e. to make packet transmission times as short as possible, the IP packets are compressed before being segmented, ciphered and transmitted. Compression means that the packet is processed and, for example, repeated sequences of zeros and ones are minimized. The compression must reduce the number of bits to be transmitted but still guarantee that the original bit sequence can be reconstructed. This function is carried out by the SNDCP layer, described in Chapter 3, before the IP packet is segmented.

2.4.2.6 Ciphering

The ciphering function in digital systems involves the combination of an information bit sequence with another bit sequence known only to the sender and the receiver (the exact procedure will be covered in Chapter 4). The goal is to ensure that the transmission is secure, i.e. the privacy of the subscriber is guaranteed and signalling information is protected. The ciphering function in GPRS is carried out by the LLC layer mentioned in Section 2.4.2.4, i.e. IP packets are compressed and segmented into LLC frames which are ciphered before being segmented into radio blocks.

2.4.2.7 Tunnelling

When an application on a mobile device sends an IP packet, the header attached to the packet contains the IP address of the MS and the IP address of the destination, e.g. an Internet or intranet address. This packet is sent across the RSS and arrives at the SGSN. The packet must now be sent from the SGSN to the GGSN via the Gn interface. A problem arises here because the Gn interface is realized using an IP network, i.e. the SGSN has an IP address and the GGSN has an IP address, which means that the IP packet from the mobile must be encapsulated in another IP packet for transport across the Gn interface – i.e. IP over IP. The problem arises because the packets from many mobiles are being sent along the same path so an extra protocol is required to differentiate between packets from different mobiles and also between different applications running on the same mobile. The GPRS tunnelling protocol defines the entry and exit points on the Gn interface, i.e. the point of encapsulation to the point of decapsulation. A tunnel is a two-way point-to-point path identified by a tunnel identifier (TID) which defines only the end points.

2.4.2.8 Domain Name Server

When an MS requests access to a data network, it sends the access point name (APN, see Chapter 4 and Chapter 6 for further details) of the requested network – e.g. intranet.tfk.de – to the SGSN. The SGSN does not know what to do with this APN, so sends a request to the domain name server (DNS), which is able to resolve this logical name into the IP address of the GGSN directly connected to the required data network. In this way the SGSN is able to contact the right GGSN.

2.4.3 Mobility Management

Mobility management (MM) has, of course, been a necessary aspect of all mobile networks prior to GPRS, but there are some innovative features in GPRS mobility management

(GMM) functionality which differ from those in circuit switched networks. These will be discussed in the following. An integration of classical GSM CS mobility management with the GPRS functionality is both possible and useful as it reduces signalling traffic on the air interface and across the network. This so called common mobility management (CMM) requires the implementation of the Gs interface from an SGSN to an MSC/VLR in order to allow signalling exchanges between the network elements responsible for MM. If this interface is not implemented, the MS makes location updates for GSM and routing area updates for GPRS separately. If the Gs interface is implemented, the MS only needs to make a single update and both the GSM and GPRS location registers will be updated.

2.4.3.1 Mobility Management States

A GPRS mobile station (MS) can be in one of three different states (Figure 2.18): idle, standby, ready. In the idle state, the mobile station is not known to the GPRS network at all. If the GPRS MS is also a mobile phone, i.e. is not a dedicated GPRS device, then it will be known to the VLR and will be making location updates while switched on. The GPRS network does not hold any location information about the MS in idle state and no paging requests nor data can be sent.

In the standby state the MS and the SGSN have established a mobility management context, i.e. the location (routing area) of the MS is stored in the SLR and the MS performs routing area updates when it passes routing area borders. Thus it can be paged for voice and data calls if necessary. In this state the MS may activate or deactivate a PDP context which is a prerequisite for actually transmitting data.

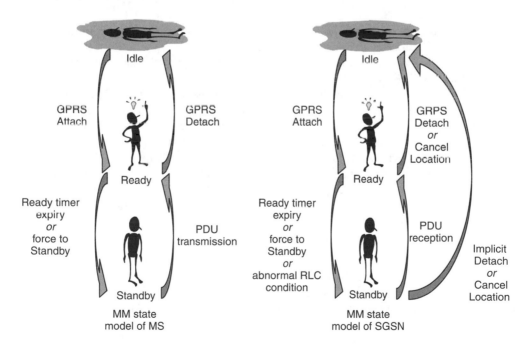

Figure 2.18 Mobility management states

If the MS is paged and responds, it will switch to the ready state. Also in the SGSN the mobility management context will be changed to the ready state. In the ready state, the exact cell location of the MS is known (comparable to an active voice call). Because of the discontinuous nature of GPRS traffic, the cell selection and reselection is usually performed by the MS, i.e. the MS monitors the surrounding base stations and decides for itself which one to use. The requests from the MS to make use of a particular cell may be accepted, or the SGSN may instruct the MS to use a different cell. The MS must be in the ready state in order to activate a PDP context to send and receive packet data. However, being in the ready state is not equivalent to having radio resources allocated to the MS, nor does it mean that a PDP context has to be active.

2.4.3.2 Transitions between Idle, Standby and Ready State

The transition from one state to another depends on the current state. Keep in mind that both the MS and the SGSN have to change to the corresponding state, and that there are some differences in how the transition may be initiated.

The transition from idle to ready state is known as the GPRS attach procedure. The MS requests access to the network, an MM context is established for its particular IMSI and the cell global identifier CGI of the selected cell is stored in the SGSN.

The standby state is usually reached by expiry of the ready timer in the SGSN. This will occur if the MS has neither sent nor received any data within the period of time defined by the timer – a typical value being 40 seconds.

In order to return to the ready state, the MS has to send any data or signalling information; this might be initiated by a page call of the SGSN. The SGSN changes the MS's state to ready only when receiving this data or signalling information, not when sending a paging request.

The transitions from ready to idle state or from standby to idle state are known as the GPRS detach procedure. Such a transition is usually initiated by a GPRS detach request from the mobile. Then the MS is stored as idle in the SGSN, which means for the network that this MS is GPRS disconnected.

There is also a special anonymous access mobility management (AA MM, Figure 2.19) for certain PDP contexts which are handled without storing any user data. In case of an AA MM, only the states ready or idle exist, i.e. if there are no PDUs to be transmitted, no data about the user are stored at all. In this case the transition from idle to ready is not called GPRS attach, but is known as 'AA PDP context activation' and the reverse is 'AA PDP context deactivation'. The expiry of the ready timer will cause a fallback to idle state and all data concerning this user will be deleted. An MS and an SGSN may have several anonymous and/or non anonymous PDP contexts at the same time.

2.4.3.3 Common Mobility Management

As previously mentioned, the common mobility management (CMM) can only be supported if the optional Gs interface between SGSN and MSC has also been implemented (see Figure 2.20). Depending on whether CMM is used or not, a network is said to be operating in a certain network operation mode. There are three network operation modes:

- A network element featuring the Gs interface that supports CMM is operating in the so called network operation mode one (NOM I). In this mode the MS will only have to

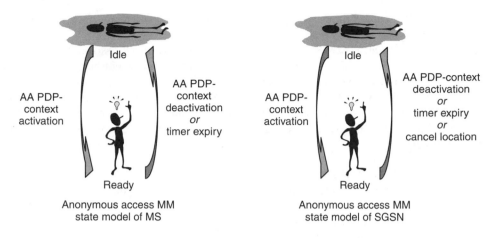

Figure 2.19   Transitions in anonymous access MM state

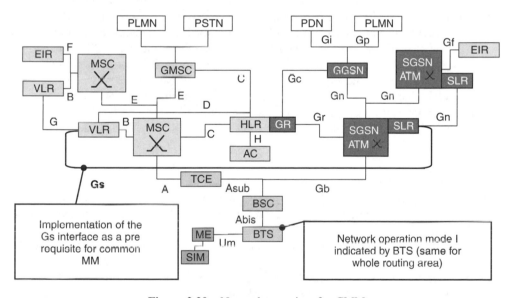

Figure 2.20   Network overview for CMM

monitor one single paging channel (GSM paging channel PCH or GPRS packet paging channel PPCH, see Chapter 5) for CS and PS calls. During a data transfer, the MS will receive paging messages via the assigned packet data channel, i.e. it does not need to monitor a separate paging channel.

- In network operation mode II (NOM II) the MS has to monitor a single paging channel as well (GSM PCH only). Unlike NOM I, the MS continues to listen to the GSM PCH during a data transfer. Paging between SGSN and MSC will not be coordinated.
- In network operation mode III (NOM III) an MS shall monitor the GPRS packet paging channel (PPCH) and the GSM paging channel (PCH).

NOM I, II or III will be indicated by the cell on the broadcast channel and should be consistent throughout a routing area.

If CMM is used, the SGSN/SLR and the MSC/VLR have to be associated with each other which is done by simply storing each others identity (SS7 identifier). There is not necessarily one SGSN for every MSC, so there are usually many VLRs associated with one SLR, especially when GPRS is first implemented in a network. Whenever an MS enters the SLR or VLR area, and the MS is both IMSI attached for voice calls and GPRS attached for data calls, the association supports the following functions in order to save radio resources:

- combined GPRS/IMSI attach and detach;
- coordinated location and routing area updates (location area update forwarded by the SGSN to the VLR);
- paging for CS connections via the SGSN;
- alert and identification procedures.

The association is initiated by the SGSN whenever a mobile requests a combined GPRS/ IMSI attach or an additional GPRS attach, or one of several combined LA/RA update cases (this includes changing to a part of the network using network operation mode I from a NOM II or III part of the network), but never updated during a CS connection.

A class 'A' terminal equipment makes RA updates during CS connection but no LA updates (LA updates are never performed during a voice call as the MS cell is always known due to handovers). A class 'B' terminal equipment does not do any updates during a CS connection as it can not provide simultaneous GSM and GPRS functionality (see Chapter 7). The association is removed on either IMSI or GPRS detach as it is no longer needed.

2.4.4 Radio Resource Management

The radio resource (RR) management procedures include the functions related to the management of the common transmission resources, e.g. the physical channels and the signalling channels.

In general, purpose of these RR procedures is to establish, maintain and release RR connections that allow a point-to-point dialogue between the network and a mobile station. This includes the cell selection/reselection and the handover procedures, which will all be discussed below. In addition to standard GPRS, the so-called E-GPRS must also be considered. E-GPRS refers to GPRS combined with EDGE (enhanced data rates for GSM evolution). EDGE is a new modulation scheme, developed to raise the net bit rate offered by one timeslot to 59.2 kbit s^{-1}. An E-GPRS mobile station has to comply with GPRS standards, as well as with E-GPRS standards which include new radio access protocol features and new modulation and coding schemes, described in more detail in Chapter 5. Regardless of whether an MS uses GPRS or E-GPRS, everything described in the following section holds true for both.

2.4.4.1 Dynamic Allocation of Resources in GPRS

The definition of a traffic channel is different between CS GSM and GPRS. In GSM a traffic channel (TCH) is bidirectional, i.e. refers to a time slot on the uplink *and* downlink, each with the same time slot number (TN). In GPRS however, a packet data traffic channel (PDTCH) is unidirectional, i.e. either UL or DL.

A cell-supporting GPRS may allocate resources on one or several PDTCHs (UL and/or DL). In order to support the GPRS traffic, those PDTCHs (possibly shared by several GPRS MS), are taken from the common pool of traffic channels available in the cell. The allocation of traffic channels to circuit switched services and GPRS can be done dynamically according to the 'capacity on demand' principles described below.

The common control signalling required by GPRS in the initial phase of the packet transfer is conveyed on GPRS's own packet common control channel (PCCCH) when implemented, or on GSM's common control channel (CCCH). This allows the operator to have capacity allocated specifically for GPRS in the cell only when a packet is to be transferred. Nevertheless the operator can also allocate packet data channels permanently. This is the case for early GPRS networks because of some restrictions in the base station system equipment and the mobile stations (these limitations will be covered later).

Master/Slave Concept

In GPRS, the physical channels are referred to as packet data channels (PDCH). At least one (PDCH) acting as a master, may accommodate the optional packet common control channels (PCCCH) that carry all the necessary control signalling for initiating packet transfer. If the PCCCH is not implemented (the standard case in early GPRS networks), that signalling is carried by the existing CCCH, which has to be present for the GSM circuit switched traffic anyway. Since the PCCCHs do not use much of the line capacity of a PDCH, there will also be user data and dedicated signalling (i.e. packet data traffic channel PDTCH and packet access control channel PACCH) on this PDCH. Other PDCHs, acting as slaves (Figure 2.21), are used for user data transfer and for dedicated signalling only.

Capacity on Demand

The GPRS does not require permanently allocated PDCHs. The allocation of capacity for actual packet transfers can be based on the demand of the subscribers, which is referred to here as the 'capacity on demand' principle (Figure 2.22). The operator can, as well, decide to dedicate permanently or temporarily some physical resources (i.e. PDCHs) for GPRS traffic only. In early GPRS networks this is usually the case for one reason in particular: power control. In GSM the power control is continuous during a circuit switched connection. The BTS measures the signal strength received from the mobile and then signals the mobile to either increase, maintain or decrease its transmission power accordingly. This is the so called 'closed loop' power control, in which sending, receiving, measuring and regulating together build a closed loop.

In GPRS there is no continuous transmission, so no continuous measurement is possible, hence the mobile station can only estimate its required transmission power level, based on the strength of the signal it receives from the BTS. This is so-called 'open loop' power control – no measurements are taken and no feedback is given. For such a system, based on estimates to be workable, the MS requires a reference on which the estimates can be based. Such a reference is provided by the carrier which broadcasts the BCCH, as this is the only carrier that is always transmitted at full power. For this reason, the time slots on this carrier will be permanently allocated to GPRS in early GPRS networks, as long as the MS and BSS are not capable of a more sophisticated power control procedures.

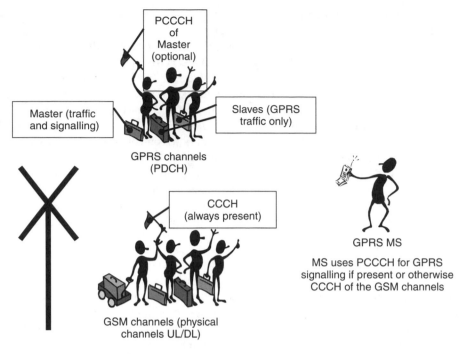

Figure 2.21   Master/slave concept

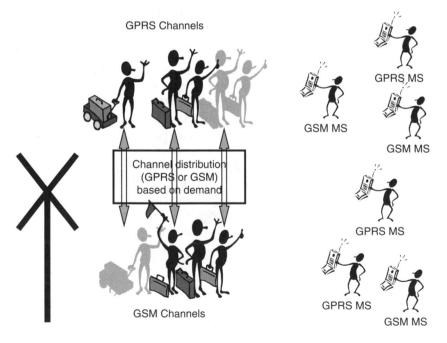

Figure 2.22   Possible combinations for capacity on demand

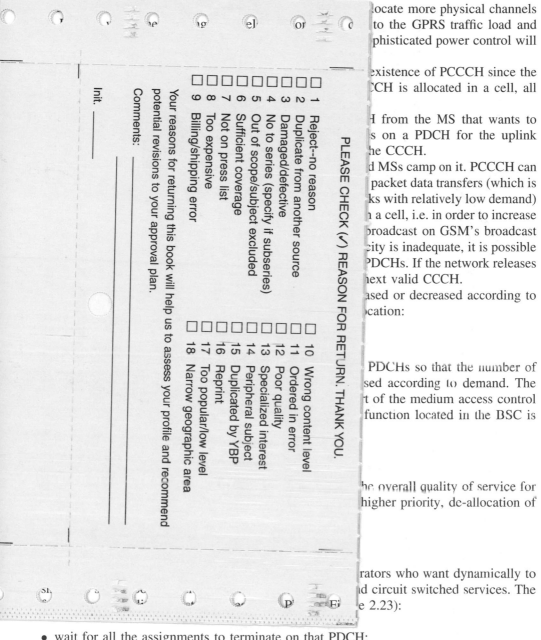

...locate more physical channels to the GPRS traffic load and ...phisticated power control will

...existence of PCCCH since the ...CCH is allocated in a cell, all

...H from the MS that wants to ...s on a PDCH for the uplink ...he CCCH.
...d MSs camp on it. PCCCH can ... packet data transfers (which is ...ks with relatively low demand) ...n a cell, i.e. in order to increase ...broadcast on GSM's broadcast ...city is inadequate, it is possible ...PDCHs. If the network releases ...next valid CCCH.
...ased or decreased according to ...ocation:

...PDCHs so that the number of ...sed according to demand. The ...t of the medium access control ...function located in the BSC is

...he overall quality of service for ...higher priority, de-allocation of

...rators who want dynamically to ...d circuit switched services. The ...e 2.23):

- wait for all the assignments to terminate on that PDCH;
- individually notify all the users that have assignment on that PDCH;
- broadcast the notification about de-allocation.

The first option does not incur any extra signalling load but could potentially take a long time. Packet uplink assignment and packet downlink assignment messages can be used to realize the second option. The network side has to send such notifications on the packet

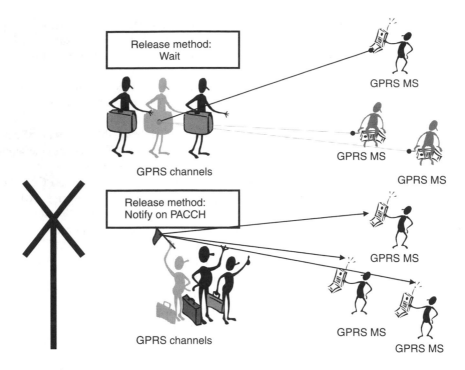

Figure 2.23   Network overview of different release methods

associate control channel (PACCH) individually to each affected MS. This method is very reliable, but slow and resource inefficient.

The 'quick and easy' approach is to broadcast the packet data channel release on all the PDCHs on the same carrier as the PDCH to be released. All MSs monitor the possible occurrences of PACCH on one channel and should capture such notification. In practice, a combination of all the methods can be used.

The situation may occur in which an MS remains unaware of the released PDCH. In such a case, the ·MS may cause some interference when wrongly assuming that the decoded uplink state flag denotes that it has been allocated the use of the following uplink block period. This 3-bit uplink state flag (USF) is the means by which up to eight mobile stations are able to share an uplink time slot (if the PDCH contains a PCCCH, one USF is reserved for the packet random access channel (PRACH), thus only seven mobile stations can be addressed and share the resource). Each MS will wait, until its personal USF is sent from the BTS and then react accordingly, but after not getting proper response from the network, the MS itself would release the RLC connection.

2.4.4.2   Radio Resource RR Operation Modes

Radio resource (RR) management procedures are characterized by two different RR operating modes. Each mode describes a certain amount of allocated functionality and information. RR procedures and RR operating modes are specified in GSM 04.07.

Packet Idle Mode

In packet idle mode, no temporary block flow exists. The temporary block flow (TBF) is a temporary physical connection used to support the unidirectional transfer of packet data units. Upper layers can require the transfer of a LLC PDU which, implicitly, may trigger the establishment of TBF and thus the transition to packet transfer mode.

In packet idle mode, the MS monitors the packet broadcast control channel (PBCCH) and the paging subchannel for the paging group the MS belongs to (*Note*: in GPRS it is possible to implement more than one paging channel so a mobile can be instructed to monitor a certain group of paging channels). As mentioned earlier, if PCCCH is not present in the cell then the mobile station monitors the BCCH and the relevant paging subchannels.

Packet Transfer Mode

In packet transfer mode, the mobile station is allocated radio resource providing a temporary block flow on one or more physical channels. Continuous transfer of one or more LLC PDUs is possible. Concurrent TBFs may (but do not have to) be established in opposite directions (as defined above, TBFs are unidirectional and do not need a TBF in the opposite direction). When selecting a new cell, the mobile station leaves the packet transfer mode, enters the packet idle mode where it switches to the new cell, reads the system information and may then return to packet transfer mode in the new cell.

While operating in packet transfer mode, a mobile station belonging to GPRS MS Class A may simultaneously enter the different RR service modes defined in GSM 04.08. A mobile station belonging to either GPRS MS Class B or C leaves both packet idle mode and packet transfer mode before entering dedicated mode for a CS call.

Dual Transfer Mode

The packet transfer mode is not applicable to MSs supporting dual transfer mode (DTM), which is defined since release 99. This mode is a subset of the terminal equipment Class A mode of operation (see Chapter 7 for a more detailed description of equipment classes) and is optional for GPRS or E-GPRS mobile stations as well as for the network. In this mode, the MS has an ongoing RR connection and is allocated a temporary block flow on one or more physical channels. Continuous transfer of one ore more LLC PDUs is possible. Concurrent TBFs may be established in opposite directions.

While in dual transfer mode the MS performs all the tasks of dedicated mode. In addition, upper layers can require:

- The release of all packet resources, which triggers the transition to dedicated CS mode,
- The release of the RR resources, which triggers the transition to idle mode and packet idle mode.

When handing over to a new cell, the MS leaves the dual transfer mode, enters the dedicated mode where it switches to the new cell, may read the system information messages sent on the SACCH, and may then enter dual transfer mode in the new cell.

It is clear from the above that there is a difference between a 'normal' Class A device and a Class A device that supports dual transfer mode. The definition of a Class A device assumes a total independence between the circuit switched and the packet switched side. As a result, such a Class A device has to be able to operate on two different frequencies at the same time in case the CS radio resources are allocated on a different carrier to the packet resources. The development of such a device is very complicated. The dual transfer mode offers a solution for this problem by allowing the transfer of packet switched data and signalling via a timeslot set up for a CS connection or other time slots on the same frequency.

Transition between RR Operation Modes

The RR operation modes, and therefore the transition between them, are dependent on the current operation mode as well as the terminal equipment operation mode (Class A, B or C). Class C mobile stations can only be attached to either GSM or GPRS. So they can only change from idle to dedicated mode and back when GSM attached (by establishing or releasing an RR state), or from packet idle to packet transfer mode and back (by temporary block flow establishment or release) when GPRS attached. The user has to change manually between GSM and GPRS attach.

For a Class A mobile station that does not support dual transfer mode, there are four possible states. The MS can be regarded as a combination of two devices, each with two possible RR states:

- On the circuit switched part of the device: idle mode and dedicated mode;
- On the GPRS part of the device: packet idle mode and packet transfer mode.

This makes a Class A MS much more versatile as it is able to swap back and forth between combinations of the operation modes, which are available only separately for the Class C MS.

For a Class B MS or a Class A mobile station that supports dual transfer mode, there are three possible RR modes:

- packet idle mode,
- packet transfer mode,
- dedicated mode.

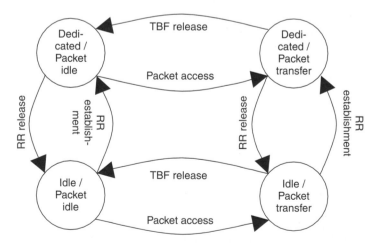

Figure 2.24 RROM and their Transitions for Class A (non DTM)

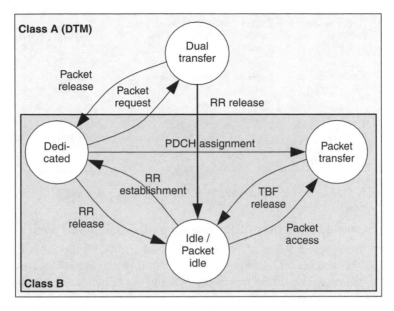

Figure 2.25   RROM and their transitions for Class B and Class A (DTM)

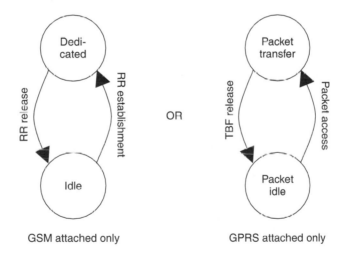

Figure 2.26   RROM and their transitions for Class C

The Class B MS can change between these three like the Class C MS, and in addition can also change from dedicated mode to packet transfer mode when assigned a PDCH. This can only be done by performing a packet request procedure while in dedicated mode. The Class A (DTM) MS, in addition to having the same capabilities as the Class B MS, can also enter the dual transfer mode.

The transitions described above are illustrated in Figures 2.24, 2.25 and 2.26.

Radio resource BSS	Packet transfer mode	Measurement report reception	No state	No state
Radio resource MS	Packet transfer mode	Packet idle mode		Packet idle mode
Mobility management NSS and MS		Ready		Standby

Figure 2.27   Correspondence between RR operating modes and MM states

2.4.4.3   Correspondence between RR Operation Modes and MM States

The mobility management states are defined in GSM 03.60. Figure 2.27 illustrates the correspondence between radio resource states and mobility management states: Each state is protected by a timer. The timers run in the MS and the network. Packet transfer mode is protected by RLC protocol timers.

3

Interfaces and Protocols

3.1 Introduction

The predominant protocol in the world of data networks is the Internet Protocol IP. The main task of IP is the networking of different network architectures and the standardization of applications, such as the exchange of e-mails between end devices or the downloading of websites.

In the fixed data networks IP is implemented on existing network architectures such as Ethernet, Frame Relay and ATM – i.e. on existing hardware – which provides the transmission paths inside the network. In the mobile network GPRS the subscriber also has a logical IP connection with an external data network – i.e. in which case IP is implemented on the GPRS network (see Figure 3.1).

Hence, within a GPRS service the mobile subscriber receives an IP address from the external IP network – i.e. he is seen as being a member of this IP network. As in a normal fixed data network, IP packets are transported between the MS and a server in the IP network. But the GPRS standard describes how these IP packets are transmitted on the radio interface and through the whole GPRS network.

Originally, the logical protocol between the MS and the external data network did not have to be IP. The older standard for packet data transfer, known as X.25, could also be used. However, since X.25 was hardly ever implemented, it has been removed from the specifications.

3.2 Layer Model

Questions often arise about the use of layered protocols. If we analyse the verbal communication between two persons, we may distinguish the elements shown in Figure 3.2.

In order that two persons are able to understand each other, they have to use the same transmission medium and the same language. The actual comprehension is done on the brain level.

In this respect the communication can be seen as a structure with several layers. Equivalent layers must understand each other. In order for concepts and ideas to be transferred between the two persons, their brains must each utilize the services of the layer below – i.e. the language. The language layer in turn makes use of the layer below – i.e.

GPRS Networks.   G. Sanders, L. Thorens, M. Reisky, O. Rulik, S. Deylitz
© 2003 John Wiley & Sons, Ltd.   ISBN: 0-470-85317-4

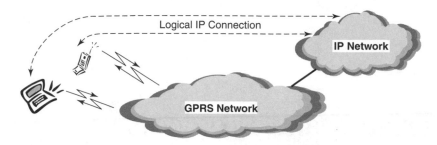

Figure 3.1 IP connections on the GPRS network

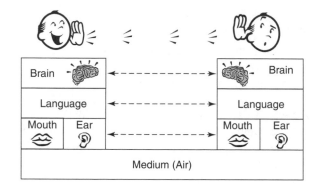

Figure 3.2 Comparison: verbal communication

the mouth. The mouth shapes the sounds produced in the vocal chords, producing modulated sound waves which propagate through the actual transmission medium – i.e. the air. The listening person follows the same process in the inverse order: the ear receives the sound waves, the language is interpreted, finally the brain receives the information 'sent' by the other person.

Considering Figure 3.3. we can see that there are two information flows. Logically, information flows horizontally between the layers. Physically however, the information

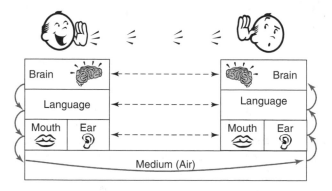

Figure 3.3 Comparison: verbal communication

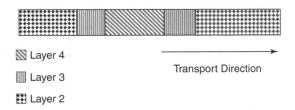

▨ Layer 4

▥ Layer 3   Transport Direction

▦ Layer 2

Figure 3.4   Structure of the messages

flows from one layer, down through the other layers, across the transmission medium and up through the layers on the receiving side. The logical communication between two equivalent layers in different systems is called the **external communication** (see dashed line in Figure 3.3). The type of external communication is defined by protocols, which define every aspect of the exchange of information between the two equivalent layers. Physically, the information to be sent must be forwarded from higher layer to lower layer downwards and on the reception side from lower layer to upper layer upwards. This forwarding – i.e. the real information flow – is called the **direct communication** (see full line in Figure 3.3).

The message being sent (by each consecutive lower layer) is embedded in further information (see Figure 3.4) which is used to communicate with the equivalent layer on the receiving side. This process is known as 'encapsulation' and results in a structure which is often likened to an onion. On the receiving side each layer reads the corresponding layer of its equivalent (external communication) and forwards the unwrapped onion to the layer above (internal communication).

The IP packets which are transported through the GPRS network between the MS and the external data network are also embedded in further information (encapsulated) in order to travel across the different interfaces. The protocols inside the GPRS network are described on the following pages.

3.3   The Names of the GPRS Interfaces

In addition to the interfaces of the classical GSM network, new interfaces for the implementation of GPRS services have had to be defined.

The following interfaces transport user and signalling data

- Gb – between BSC and SGSN
- Gi – between GGSN and external Public Data Network (PDN)
- Gn – between SGSN and GGSN (or SGSN and SGSN) within one mobile network
- Gp – same as Gn but between two different mobile networks

The following interfaces only transport signalling data

- Gc – between GGSN and HLR/GR
- Gf – between SGSN and EIR
- Gr – between SGSN and HLR/GR
- Gs – between SGSN/SLR and MSC/VLR

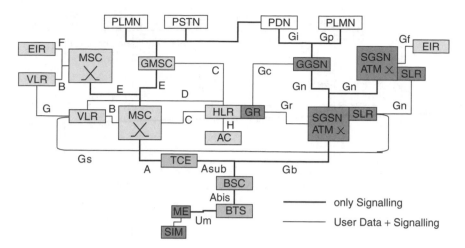

Figure 3.5   The interfaces of a GPRS network

The interface Gd (between an SGSN and an MSC connected to the SMS Centre) is defined for the transport of SMS (Short Message Service).

All the interfaces described above are shown in Figure 3.5. It should be noted that not all these interfaces are realized as direct connections between network elements. Some, such as the Gc interface, may be realized via other network elements in order to minimize the number of physical interfaces required. In the case of the GGSN to HLR interface, Gc may be realized by sending the signalling via an SGSN.

3.4   GPRS Procedures

Before data can be transferred between the MS and the external data network, some preparation is necessary to enable the transfer of IP packets through the GPRS network to take place. There are three important steps involved (see Figure 3.6):

1. The MS must be attached in the GPRS network. The procedure for this is called 'GPRS attach'. This is a logical procedure between the MS and the SGSN which takes note of the position (i.e. the 'routing area', as described later) of the MS. Storing and up-dating the position of the MS is particularly important for DL transmissions because this information enables the GPRS network to locate the MS.
2. How is the right path found for the IP packets inside the GPRS network? A connection between the MS and the GGSN must be set up, we speak of the activation of a PDP context. After that procedure each node in the GPRS network knows how it has to forward the IP packets of this MS.
3. The path between the MS and the external data network is prepared, so IP packets can be sent through the GPRS network towards the destination address.

These three procedures and the necessary protocols on each interface, are described in the following text.

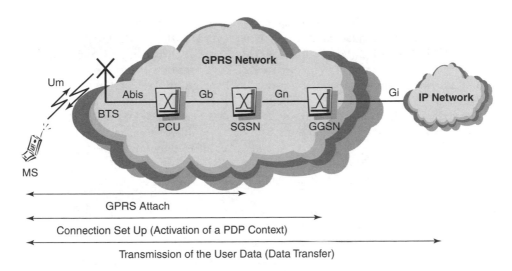

GPRS Attach

Connection Set Up (Activation of a PDP Context)

Transmission of the User Data (Data Transfer)

Figure 3.6 GPRS procedures

3.5   GPRS Attach

A subscriber requests a GPRS attach procedure when he registers with the GPRS network. This is the case when he switches on his mobile device or he explicitly activates GPRS while already GSM attached.

The result of this procedure is that the current SGSN (i.e. SLR) knows that the MS has activated GPRS. If the MS was already registered with another SGSN, the new SGSN updates the HLR such that the HLR (i.e. GR) knows the current SGSN identity. The HLR then sends the GPRS specific data of the MS to this current SGSN. This GPRS specific data corresponds to the different PDP contexts (PDP = Packet Data Protocol, see next section) which are set up for this MS. A PDP context describes the GPRS data connection of an MS. The information which describes such a context are sent from the HLR and consist of:

- the Access Point Name APN: a logical name for the desired data network – e.g. intranet.tfk.de,
- the Quality of Service QoS – priorities, delays, reliability, throughputs – for the desired application (speech, video, internet surfing, download, etc.),
- the PDP protocol: the protocol used between the MS and the external data network, usually IP (version 4 or 6),
- the permanent address (rather with IP version 6) of the MS if the MS has such an address.

Several PDP contexts can be set up for an MS such that the MS can reach several IP networks (i.e. several APNs) with different QoS. Several APNs can also be given to the same data network for different services with different QoS.

The messages concerning the GPRS attach procedure between the MS and the SGSN belong to the protocol GPRS Mobility Management GMM (see Figure 3.7). Other

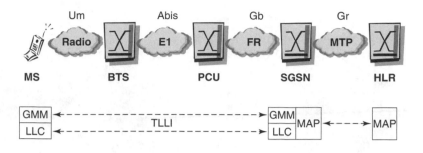

GMM: GPRS Mobility Management   E1: Link with 2048 kbit/s   MAP: Mobile Application Part
LLC: Logical Link Control   FR: Frame Relay   MTP: Message Transfer Part
TLLI: Temporary Logical Link Identifier

Figure 3.7   GPRS attach procedure

examples of GMM (see subsection 9.1) are the update of the MS's position (routing area update), switching off of the mobile device (GPRS detach), authentication, and GPRS paging.

Another important protocol between the MS and the SGSN is the Logical Link Control LLC which provides a logical connection between the MS and the SGSN. It is defined by the Data Link Connection Identifier DLCI which consists of the Temporary Logical Link Identifier TLLI and the Service Access Point Identifier (SAPI). The TLLI identifies the specific MS and the SAPI identifies the service access point (SNDCP, GMM/SM or SMS) – i.e. whether the data in the packet's payload concerns user data, signalling or SMS. In the case of a GPRS attach procedure the SAPI shows that it concerns GMM/SM (SM = Session Management, see the next section on PDP context). LLC is also responsible for ciphering the connection – i.e. a GPRS connection is ciphered between the MS and the SGSN (with circuit switched GSM, transmissions are only ciphered between the MS and the BTS).

The Gr interface between the HLR and the SGSN is based on the MAP (Mobile Application Part) protocol which is the application layer of all signalling interfaces inside the GSM core network (see Figure 3.7 on how MAP is involved in the GPRS attach procedure).

Other protocols on the lower layers are of course also defined, but they are described later.

3.6   Activation of a PDP Context

A subscriber requests the activation of a PDP context (PDP = Packet Data Protocol) when he wants to start a GPRS service. For that he must select a start page and an Access Point Name APN (e.g. intranet.tfk.de) with a certain Quality of Service QoS (see Figure 3.8 on which protocols are involved in the PDP context activation).

The result of this procedure is that the whole GPRS network knows how to route the IP packets of the GPRS service. The path from the MS up to the GGSN (actually up to the Gi interface) is well defined after the activation of the PDP context.

The messages concerning the activation of the PDP context between the MS and the SGSN belong to the protocol Session Management SM. Another example of SM is the

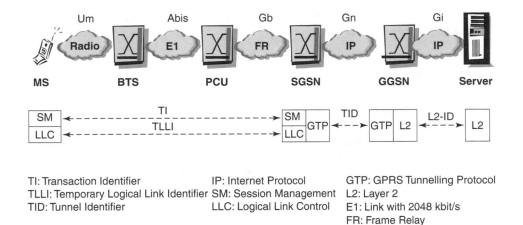

Figure 3.8   Activation of a PDP context

deactivation of the PDP context. A Transaction Identifier TI is allocated to an activated PDP context. The TI identifies the session on the SM layer.

As described previously, the Logical Link Protocol LLC enables information to be transferred between the MS and the SGSN. The logical connection between the MS and the SGSN is defined by the Data Link Connection Identifier DLCI which consists of the Temporary Logical Link Identifier TLLI and of the Service Access Point Identifier SAPI. The TLLI remains the same unless the SGSN (i.e. SLR) has allocated a new TLLI since the GPRS attach procedure. The SAPI shows that the payload of the data packet concerns GMM/SM. Furthermore, LLC is responsible for ciphering the connection between the MS and the SGSN.

The information which belongs to the PDP context is sent over the Gn interface with the GPRS Tunnelling Protocol GTP, which uniquely identifies the PDP context of a subscriber between the SGSN and the GGSN with a Tunnel Identifier TID. The TID is a combination of the International Mobile Station Identity IMSI and the Network Service Point Identifier NSAPI. The IMSI is a fixed identifier of the MS and the NSAPI identifies the application (the NSAPI is also defined on the SNDCP layer between the MS and the SGSN, see next section).

The GGSN has two additional tasks:

- it sets the right physical port (layer 2) on the Gi interface to this PDP context, such that the IP packets of this PDP context are routed to the correct external data network. The layer 2 can be Ethernet, Frame Relay or ATM (the possible implementations are the same as those of the Gn interface, see subsection 8.3).
- usually the MS has no permanent IP address (it would be possible with IP version 6), therefore the MS must receive a temporary one (as it is currently the case with IP version 4). The MS should receive such a temporary IP address from the external IP network and the GGSN is responsible for this task – i.e. the GGSN requests an address from the IP network and forwards it to the MS.

Other protocols on the lower layers are of course also defined, but they are described later.

3.7 Data Transfer

The PDP context to the APN (see Figure 3.9) selected by the MS is now activated – i.e. the logical path between the MS and the external IP network is now defined:

- The MS has received a temporary IP address from the external IP network, so the MS is seen as being a member of that IP network,
- Towards the SGSN the MS is identified by the TLLI (on the LLC layer) and the PDP context is identified by the NSAPI (on the SNDCP layer),
- Between the SGSN and the GGSN each particular MS and each of its PDP contexts is identified by the TID,
- On the Gi interface the GGSN sets the right physical port (layer 2) on the Gi interface to this PDP context.

In this way the routing of incoming or outgoing IP packets for this PDP context is completely defined: the SGSN knows that an IP packet between the GGSN and itself must be marked with a certain TID if this packet between the MS and itself is marked with the corresponding NSAPI and TLLI, and vice versa. The GGSN again knows which TIDs on one side correspond with IP addresses and the physical port on the other side. The external network (internet or intranet) is not aware of this at all. The GGSN just appears to be an external access router (from the point of view of that network).

The IP packets which are transported between the MS and the external data network are compressed and segmented between the MS and the SGSN: this is one task performed by the Sub-Network Dependent Convergence Protocol SNDCP. When data packets are transmitted between the MS and the SGSN, the SNDCP layer compresses and segments the IP packets, and when the MS or the SGSN receives these data, SNDCP re-assembles und decompresses the LLC packets to get the IP packets again.

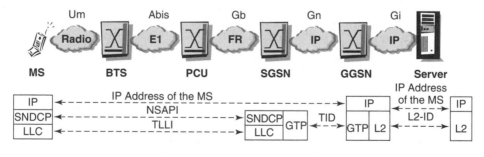

NSAPI: Network Service Access   IP: Internet Protocol   GTP: GPRS Tunnelling Protocol
       Point Identifier   SNDCP: Sub Network Dependent   L2: Layer 2
TLLI: Temporary Logical Link Identifier   Convergence Protocol   E1: Link with 2048 kbit/s
TID: Tunnel Identifier   LLC: Logical Link Control   FR: Frame Relay

Figure 3.9 Logical data transfer

As before we have the Logical Link Protocol LLC between the MS and the SGSN. The only difference is that the SAPI now shows that it concerns SNDCP (i.e. user data).

In this way the logical data transfer is completely defined by the activation of a PDP context. The physical data transfer has not yet been described and is covered in the next section.

3.8   Physical Implementation in the GPRS Network

3.8.1   The Interfaces Um and Abis

The Radio Link Control RLC (between the MS and the PCU) segments the LLC packets into smaller packets called 'radio blocks' (see subsection 8.1.2) for transmission over the radio interface and re-assembles the received radio blocks from the radio interface (and from the Abis interface) into LLC packets. In the RLC 'acknowledged mode' of operation, the RLC layer also provides the Backward Error Correction BEC procedures enabling the selective retransmission of unsuccessfully delivered radio blocks.

The MAC protocol enables several mobile stations to share a common transmission medium and also allows a mobile station to use several physical channels in parallel – i.e. use (and share) several timeslots within the TDMA frame (see subsection 8.1.1).

The physical layer of the radio interface describes the physical transmission of the digital information. Some checksums are added to the user information such that detection of errors is possible. These checksums and a part of the user information (see Section 8.1.2) are copied (actually the whole information is copied but only part of this so called 'redundancy' is transmitted, see Section 8.1.2) such that correction of the detected errors is possible. This method for securing the reliability of the transmission before sending is called Forward Error Correction FEC. As shown in Figure 3.5. the radio access in GSM is determined by narrow frequency bands of 200 kHz (Frequency Division Multiple Access – FDMA) and by timeslots (Time Division Multiple Access – TDMA). The two directions of the connection (uplink and downlink – UL and DL) are transported by different frequencies (Frequency Division Duplex – FDD).

The first layer of Abis is realized with E1 links (2048 kbit/s). See Figure 3.10 on the position of GSM radio frequency, E1 and RLC layer.

3.8.1.1   Channel Bundling and Sharing of Channels

In GSM Phase 1 and Phase 2, a subscriber receives one timeslot for UL and one timeslot for DL. These timeslots are allocated to the subscriber for the duration of the call (it is a symmetric circuit switched connection). In GPRS there are two distinct differences:

- Several timeslots on one carrier frequency may be allocated to one user – this is known as 'bundling' of timeslots. Timeslots can be bundled on the UL and the DL. The allocation of timeslots may also be asymmetric. For instance, a subscriber who wants to download some data from the internet will receive more timeslots for the DL than for the UL.
- One timeslot is not reserved exclusively for one subscriber – i.e. a timeslot may be shared by several subscribers. As a GPRS connection is packet switched, the packet transmissions from several users can be multiplexed through the same timeslots on the Um. In the case that several subscribers are sharing an UL timeslot and all of them

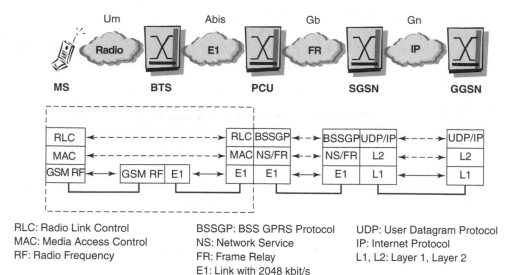

RLC: Radio Link Control   BSSGP: BSS GPRS Protocol   UDP: User Datagram Protocol
MAC: Media Access Control   NS: Network Service   IP: Internet Protocol
RF: Radio Frequency   FR: Frame Relay   L1, L2: Layer 1, Layer 2
                                 E1: Link with 2048 kbit/s

Figure 3.10   Physical implementation of the interfaces Um and Abis

want to transmit, if none of them has a higher priority than the others the PCU allows each MS in turn to transmit some data. So the mobiles transmit one after the other until each transmission is complete.

There are two parameters which are important for the allocation of the GPRS radio resources:

- the Temporary Flow Identifier TFI
- the Uplink State Flag USF

These parameters are sent to the mobile as part of the MAC protocol header and are used to give information about channel usage to the mobiles which are sharing a timeslot.

Several subscribers may share a radio channel on the DL. Therefore in each DL radio block (see next section) a TFI is necessary for determining the owner of each packet. A TFI has 5 bits, so 32 different values are possible – i.e. up to 32 subscribers may theoretically share a DL radio channel.

Several mobiles may also share a radio channel on the UL. These mobiles must be informed when it is their turn to send. Therefore, an additional parameter is sent in each DL radio block: the USF. It indicates which subscriber can send next on the relevant UL radio channel. An USF has 3 bits, so 8 different values are possible – i.e. up to 8 subscribers may theoretically share an UL timeslot. If one UL timeslot has been configured (reserved) as a PRACH (Packet Random Access Channel – used by an MS to request a connection) then USF = 111 is reserved to identify the PRACH and the other 7 values (000 to 110) remain to identify up to 7 subscribers on this UL timeslot.

3.8.1.2 Radio Blocks and Coding Schemes

The digital information which is sent over the radio interface is divided into 'radio blocks'. One radio block contains 456 bits and is the information which can be sent in a certain timeslot in four consecutive TDMA frames (see Figure 3.11). This means that if one mobile is allocated a timeslot (UL or DL) for transporting information, the mobile actually transmits four times on this timeslot, because one radio block is the smallest unit of information over the radio interface. Therefore the sharing of channels is not carried out on a 'timeslot-by-timeslot' basis, but instead on a 'radio block-by-radio block' basis. This process is shown in Figure 3.11 which shows the timeslots belonging to each user being utilized for a minimum of 4 consecutive TDMA frames. After 4 TDMA frames timeslot 2 is allocated to a different user. After another 4 TDMA frames timeslot 1 is also re-allocated and timeslots 5 and 6 are allocated to the user of timeslot 3.

Four different coding schemes CS have been defined for GPRS. Each radio block is coded using one of these coding schemes. They may be used alternatively, depending on the quality of the radio interface, but the CS can not be changed during a radio block transmission – i.e. the CS can only be changed every 4 TDMA frames (the duration of one radio block) if necessary.

CS-1 is a very reliable coding scheme as 'only' 181 user bits are sent per radio block. Additionally, there are the 3 bits of the Uplink State Flag, a Block Check Sequence of 40 bits (for detecting errors) and 4 Tail Bits. Together that makes: $181 + 3 + 40 + 4 = 228$ bits, and all of them are doubled (copied) such that we get the exact number of bits ($2 \times 228 = 456$) which a radio block can carry. With this coding scheme the information sent over the radio interface has 100 % redundancy.

CS-2 is a less reliable coding scheme because 268 user bits are sent per radio block which means there are fewer bits available in the radio block for redundancy. In addition to the user bits, there are 2 times the 3 bits of the Uplink State Flag, a Block Check Sequence of 16 bits (for detecting errors) and 4 Tail Bits. Together we have: $268 + 6 + 16 + 4 = 294$ bits, and all of them are copied such that we get 588 bits. As a radio block can only carry 456 bits, 132 bits must be discarded. This is done in a systematic way through a procedure known as 'puncturing' – i.e. CS-2 has 132 punctured bits.

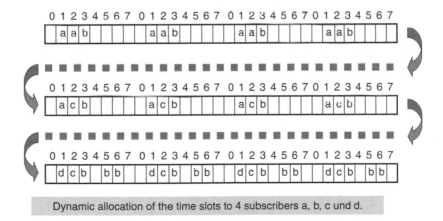

Dynamic allocation of the time slots to 4 subscribers a, b, c und d.

Figure 3.11   Radio blocks

CS-3 is an even less reliable coding scheme than CS-2 because 312 user bits are sent per radio block with correspondingly less redundancy. In addition to the user bits there are 2 times the 3 bits of the Uplink State Flag, a Block Check Sequence of 16 bits (for detecting errors) and 4 Tail Bits. Together we have: $312 + 6 + 16 + 4 = 338$ bits, and all of them are copied such that we get 676 bits. As a result CS-3 has 220 punctured bits in order to fit into a 456 bit radio block.

CS-4 is an unreliable coding scheme because 428 user bits are sent per radio block which means only error checking bits can be added. In addition to the user bits there are 4 times the 3 bits of the Uplink State Flag and a Block Check Sequence of 16 bits (for detecting errors). Together we have: $428 + 12 + 16 = 456$ bits, which is the exact number of bits in a radio block. No bits must be punctured because the information sent over the radio interface with this coding scheme contains no redundancy at all.

The four GPRS coding schemes, the bit rates and puncturing, are shown in Figure 3.12. The net data rate of each coding scheme is calculated as shown in Figure 3.13.

The 20 ms duration used in the calculation is confusing for many people. What are these 20 ms? Intuitively we would think that it corresponds to the duration of a radio block. This answer is not absolutely correct.

A TDMA frame has a duration of 4.615 ms – i.e. the transmission of 4 successive TDMA frames is 18.46 ms, so it is smaller than 20 ms. The answer to the above question becomes apparent when looking at the GPRS multiframe (see Figure 3.14).

Within GPRS, the TDMA frames are included in a larger time structure consisting of 52 successive TDMA frames: such a structure is a GPRS multiframe. The first 12 TDMA frames are for user data and they are divided in 3 radio blocks. After these 3

	Duration of radio block	Net number of bits	Preco- ded USF	BCS	Tail bits	Number of coded bits	Punc- tured bits	Net data rate
CS-1	20 ms	181	3	40	4	456	0	9,05 kbit/s
CS-2	20 ms	268	6	16	4	588	132	13,4 kbit/s
CS-3	20 ms	312	6	16	4	676	220	15,6 kbit/s
CS-4	20 ms	428	12	16	0	456	0	21,4 kbit/s

Figure 3.12 The four GPRS coding schemes

$$\text{Net Data Rate} = \frac{\text{Net Number of Bits}}{\text{Duration}}$$

$$\text{Example for CS-1: } 9,05 \text{ kbit/s} = \frac{181 \text{ Bits}}{20 \text{ ms}}$$

Figure 3.13 Net data rate

B0-B11: Radio Blocks 0-11 X: For Signalling

Figure 3.14 GPRS multiframe

radio blocks a single TDMA frame is reserved for signalling (e.g. for power control or timing advance). When we repeat this structure of 12 TDMA frames for user data and one TDMA frame for signalling and we get a whole GPRS multiframe. We can notice that for a fixed timeslot number not all successive timeslots are usable for user data: of 13 successive timeslots only 12 are usable.

This timeslot which is reserved for signalling reduces the data rate – i.e. it increases the duration of the transmission: $\mathbf{18.46 \times 13/12 = 20\ ms}$. Because of the additional timeslot for signalling an overall duration of 20 ms must be considered for the transmission of a radio block.

3.8.2   The Interface Gb

On the Gb interface, the PCU only interprets the BSSGP information (see subsection 8.2.2) which is used in the DL direction by RLC/MAC and derived from RLC/MAC in the UL direction.

According to the specifications Gb is based on the Frame Relay technology which is described in the following subsection 8.2.1. In order to provide an end-to-end communication between the PCU and the SGSN, irrespective of the exact implementation of the Gb interface, a Network Service Virtual Connection (NS-VC) is configured.

Physically, the information is transported on an E1 link (see Figure 3.15).

3.8.2.1   Frame Relay

Frame Relay is a packet oriented transmission technology for layer 2, which makes it possible to connect local networks over long distances. Frame Relay uses existing PDH (Plesiochronous Digital Hierarchy) transmission technologies (E1 at 2 Mbit/s and E3 at 34 Mbit/s), therefore the integration in existing transmission networks is quite straight forward.

Frame Relay is connection oriented and the packets can have a variable length, which depends on the length of the user data. The data which is transmitted with Frame Relay, for example an IP packet, is packed in a Frame Relay packet. Each packet receives an address which enables the network to find the destination. A packet with such an address is known as an 'FR frame'. The frames travel on an existing connection through the switches of the Frame Relay network – i.e. this means that first a connection must be set up.

The fields of an FR frame are illustrated in Figure 3.16. which clearly shows the leading and trailing frame alignment bit-sequences, and the variable length data field which gives FR its flexibility.

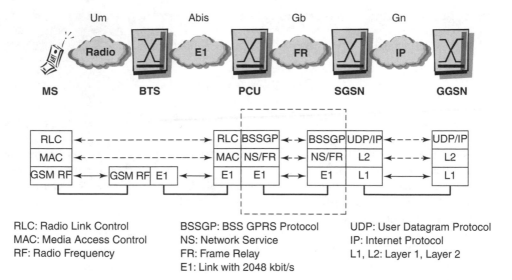

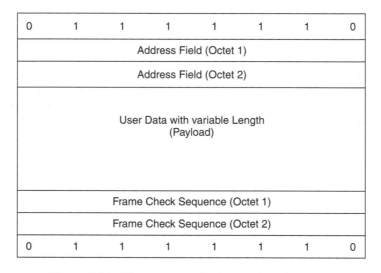

RLC: Radio Link Control   BSSGP: BSS GPRS Protocol   UDP: User Datagram Protocol
MAC: Media Access Control   NS: Network Service   IP: Internet Protocol
RF: Radio Frequency   FR: Frame Relay   L1, L2: Layer 1, Layer 2
                                 E1: Link with 2048 kbit/s

Figure 3.15   Physical implementation of the interface Gb

0	1	1	1	1	1	1	0
Address Field (Octet 1)							
Address Field (Octet 2)							
User Data with variable Length (Payload)							
Frame Check Sequence (Octet 1)							
Frame Check Sequence (Octet 2)							
0	1	1	1	1	1	1	0

Figure 3.16   The structure of a frame relay (FR) frame

The Frame Relay switches themselves only perform error detection and discard faulty frames. To maintain a low error rate, the Frame Relay elements should be connected to each other by links with a high transmission reliability.

3.8.2.2   BSSGP

BSSGP is the protocol which transports 'subscriber relevant' information over the Gb interface (see Figure 3.11). This information may for instance concern a paging or a random access. Examples of parameters which may be transported by BSSGP are the TLLI (which identifies the subscriber), the QoS parameters (for the subscriber's application such that the PCU is able to allocate an adequate amount of resources over the Um and Abis interfaces) and the Routing Area (such that the PCU knows which cells the subscriber must be paged in).

3.8.3 The Interface Gn

The GPRS Core Network is an IP (known as Gn – see Figure 3.17) based network implemented with UDP as the transport protocol. UDP is a connectionless protocol offering port addressing and header checksums, but no other functionality other than that provided by IP itself – i.e. there is no error checking or flow control.

The IP layer corresponds to the network layer (layer 3). While hardware addresses are used in the data link layer (layer 2) in order that data may be exchanged between two devices, these hardware addresses are translated into IP addresses in the IP layer. The hardware addresses have a structure which is dependent on the network technology, while IP addresses have a general structure which is independent of the hardware. So, for example, it is possible to send data from a device with an Ethernet interface to a device with an ATM interface. The data transmission is 'connectionless' which means that there is no end-to-end connection set up before the data is transmitted, as there is for example, with a phone call. Each IP packet is individually sent (routed) through the network and each router on the path between the two end devices reads the IP address of the destination from the packet header. The IP address of each node can only exist once within a closed network as it is the case of the SGSNs and the GGSNs within the core network of a mobile network.

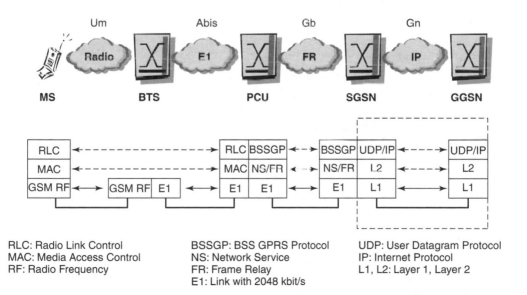

RLC: Radio Link Control   BSSGP: BSS GPRS Protocol   UDP: User Datagram Protocol
MAC: Media Access Control   NS: Network Service   IP: Internet Protocol
RF: Radio Frequency   FR: Frame Relay   L1, L2: Layer 1, Layer 2
                                 E1: Link with 2048 kbit/s

Figure 3.17   Physical implementation of the interface Gn

	Setup Option 1	Setup Option 2	Setup Option 3
Network layer	IP	IP	IP
Data link layer	ATM	Frame Relay	Ethernet
Physical layer	STM-1 (optical fibre, 155 Mbit/s)	E3 (34 Mbit/s)	10/100BaseT (10/100 Mbit/s)

Figure 3.18   The Gn interface

There are several possibilities for layer 2 and layer 1 which depend on the preferences of the operator. The different transmission technologies on layer 2 are Ethernet, Frame Relay and ATM. The different implementations are summarized in Figure 3.18.

These are the most widely used solutions. But ATM on E3 (34 Mbit/s) or E1 (2 Mbit/s) and Frame Relay on E1 are also possible.

Frame Relay is rarely used. If the distance between the SGSN and the GGSN is small Ethernet is usually selected. This technology is easy to implement and cheap. Another technology which is also often used is ATM. ATM uses E1, E3 or STM-1 on layer 1 and can therefore be integrated in existing transmission networks. The distances between the nodes can be large. An additional advantage of ATM is that a lot of parameters for Quality of Service on layer 2 can be configured, although this advantage is not utilized in current IP networks.

3.8.4   The Interface Gp

The Gp interface is necessary in case a subscriber wants to reach a data network which is only connected to a GGSN in another mobile network – i.e. an interface between the current SGSN and the GGSN of the other mobile network must be implemented. The protocols of the Gp interface are the same as the ones for the Gn interface. In addition to the functionality of the Gn interface, the Gp interface offers extra security functionality which is necessary for connections between different networks. This security functionality is based on agreements between both network operators.

3.9   GPRS Signalling

3.9.1   GPRS Mobility Management

In order for a record of the current position of a subscriber to be maintained in the network, some data about the position has to be exchanged quite regularly between the MS and the network – i.e. the location records must be regularly updated. If an update was performed every time the MS changed cells, a very large amount of signalling would result. If the position of a subscriber was not updated, it would be necessary to search for the subscriber in the whole mobile network in case of an incoming call (paging). Again, a large amount of signalling would be the consequence. A compromise between these two extreme cases must be found for every network – i.e. the solution which minimizes the amount of signalling required for location updates and paging.

As shown in Figure 3.19, the signalling for location updates can be plotted against the number of cells in a location area. On the same graph the signalling for paging can be plotted against the number of cells in a location area. The point where the two curves cross gives the size of a location area (i.e. the lowest number of cells) which requires the least amount of signalling in total. The higher amount of paging involved in GPRS leads to a smaller number of cells in the groups.

Having defined the optimum size for the location areas, the network operators divide their networks logically into groups of cells and each group is given a unique ID which is broadcast on the BCCH (Broadcast Control Channel – see 'Chapter 1 – Introduction' and 'Chapter 5 – Radio Sub-System Modifications') of each cell in the group. The MS decodes this ID as it moves from cell to cell. If the ID changes, the MS knows that it has moved to a different group and must perform a location update – i.e. it signals the network that it has changed areas and the ID of the new area is stored in the network. In GSM this is called a Location Update and in GPRS a Routing Area Update. When a subscriber receives a call, he will be paged in all the cells belonging to this group. Such a group of cells is called a Location Area for the circuit switched domain and the current Location Area ID (LAI) for each mobile is stored in the VLR. In the packet switched domain, a group of cells is called a Routing Area and the current Routing Area ID (RAI) for each mobile is stored in the SLR (SGSN Location Register).

The information about the position of a subscriber is managed hierarchically. The first two digits of a phone number (MSISDN) after the network prefix number, correspond to the identity of the HLR of the subscriber. A mobile network may have one or several HLR's. The HLR knows the current Visitor Location Register (VLR) for all of its subscribers. Every MSC has its own VLR, and the current VLR knows the current Location

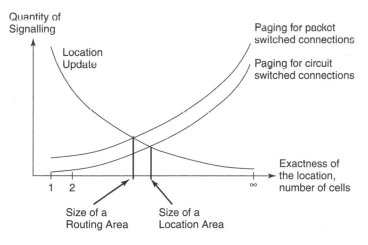

Figure 3.19   Compromise between location updates and paging

Area of the subscriber. When the subscriber moves from one cell to another inside a Location Area, no Location Updated is initiated. If he moves to a new Location Area, the VLR is informed about this change. If he moves to a new VLR area, the HLR is also informed about this change and the record in the previous VLR is erased. In this way the amount of signalling for Location Updates is minimized. In case of an incoming circuit switched call to the subscriber the GMSC contacts the subscriber's HLR which will then contact the current VLR to get the network address of the current MSC. The GMSC contacts the current MSC which asks the VLR for the current Location Area. The MSC then initiates a paging operation to the subscriber's MS in all the cells belonging to the current Location Area. When the MS responds to the paging, the subscriber's current cell is identified and the call can be switched through. In this way the amount of signalling for paging is minimized.

This hierarchical approach is illustrated in Figure 3.20, which shows the groups of cells belonging to different Location Areas, and various Location Areas belonging to different MSC/VLR's.

In case of an incoming packet switched connection, the quantity of signalling may be larger because it is possible that a subscriber must be paged several times during a download. As a consequence the number of cells in a group for GPRS is smaller than the number of cells in a Location Area. The reason for this can be clearly seen in Figure 3.19. The curve for GPRS paging crosses the updates curve earlier, which means the compromise between Updates and Paging requires smaller groups of cells. This smaller group of cells is called a Routing Area and the RAI of each subscriber is stored in the SLR (see Figure 3.21).

Hence a Routing Area is typically smaller than a Location Area, although in some networks the Routing Areas are the same as the Location Areas – e.g. one operator in the Czech Republic had Location Areas = Routing Areas = 100 cells; one German operator had Location Areas = 9 cells but Routing Areas = 3 cells. For GMM it is important that

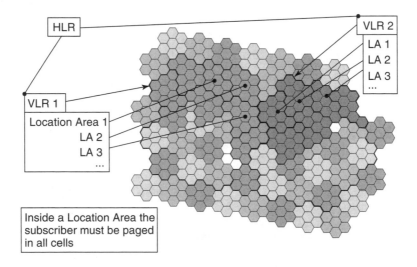

Figure 3.20   Hierarchical localization within GSM

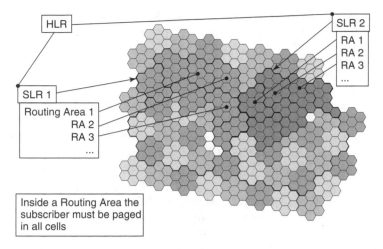

Figure 3.21   Hierarchical localization within GPRS

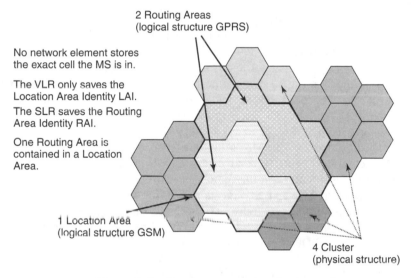

Figure 3.22   Routing area vs. location area

every Routing Area lies entirely within one Location Area – i.e. a Location Area contains one or several Routing Areas and no overlaps are allowed – as shown in Figure 3.22.

When the subscriber leaves a Location Area, he also moves to a new Routing Area. Therefore an update of both the Location Area in the VLR and of the Routing Area in the SLR are necessary. If the additional Gs interface between the MSC/VLR and the SGSN/SLR is implemented, the MS does not contact both SLR and VLR. In order to minimize the quantity of signalling over the radio interface the MS only updates the SLR. The SLR saves the new Routing Area Identity (RAI) and contacts the VLR via the Gs interface, the VLR then saves the new Location Area Identity (LAI). This procedure is known as 'Common Mobility Management' (CMM).

For more detailed information about the GPRS Mobility Management procedures, see Chapter 4 'GPRS Procedures'.

3.9.2 Session Management

Session Management (SM) for the transmission of data to and from the subscriber, is one of the tasks performed by the SGSN. The SM status of a subscriber is either *active* or *inactive* – i.e. data transmission to or from this subscriber is either possible or not. In order for the SM status to become active, the so called 'PDP context' (PDP = Packet Data Protocol) must be activated. The PDP context is determined in the MS, in the SGSN and in the GGSN and contains all relevant parameters for the desired data transmission:

- Quality of Service (QoS) – priorities, delays, reliabilities, throughputs – for the desired application (speech, video, internet surfing, download, etc.)
- IP address for the MS – Temporary (with IP version 4) or fixed (with IP version 6)
- Access Point Name (APN) which is the logical name for the desired data network – from this name, the IP address of the GGSN is determined in a Domain Name Server (DNS).

If a PDP context has been activated and if the subscriber does not transmit or receive anything, the PDP context remains active only for a specific period of time. This period of time is determined by the operator and is controlled by a timer. The period is typically around one hour and is restarted every time the MS sends or receives some data. This means that a GPRS subscriber keeps his temporary IP address only until the timer expires, after which the MS goes into PDP inactive state.

The following two Figures 3.23 and 3.24 show how the Session Management for a particular application works. The MS is first switched on and this state is stored in the SGSN. If the subscriber wants to send data, the MS activates a PDP context for the desired application. Resources are then allocated for the transmission (RAB for Radio Access Bearer between MS and SGSN and CNB for Core Network Bearer between SGSN and GGSN). After the data transmission, the resources are released.

Later, if the MS again wants to send data and the timer has not expired, the MS does not need to activate the PDP context again.

If the GGSN wants to send data to the MS, the MS must first be located (paged) in the Routing Area which is currently stored in the SLR. When the MS receives the paging request it must respond and, if the PDP context is in the inactive state, activate a PDP context in order to receive the data.

Finally, at the end of a session, the MS is switched off (this state is saved in the SGSN) and the PDP context is deactivated.

For more detailed information about the Session Management procedures, see Chapter 4 'GPRS Procedures'.

3.9.3 The Other Interfaces in the Core Network

3.9.3.1 The Interfaces Gd, Gf, Gr, Gs

These interfaces are based on SS7 (Signalling System number 7) like within circuit switched GSM core network.

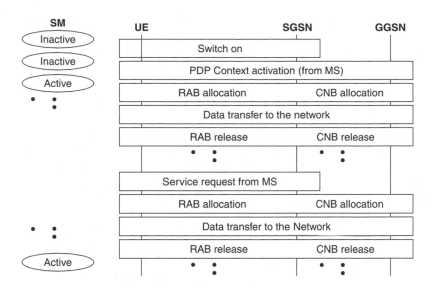

Figure 3.23   Session management for a data transmission

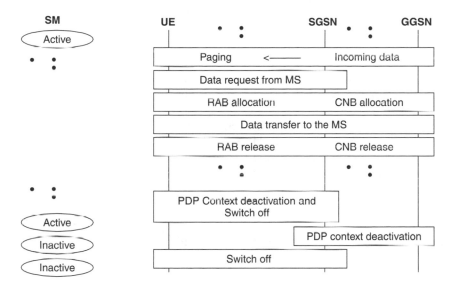

Figure 3.24   Session management for a data transmission (after paging)

The Gs interface between the MSC/VLR and the SGSN/SLR (as shown in Figure 3.25) is necessary for Common Mobility Management (CMM) tasks as described in the subsection 9.1. The application layer BSSAP+ is a subset of the application layer BSSAP of the GSM A interface between the MSC/VLR and the BSC.

The Gr interface transports all signalling information between the HLR and the SGSN/SLR. The Gf interface may be used if the optional network element EIR (Equipment Identity Register) is implemented.

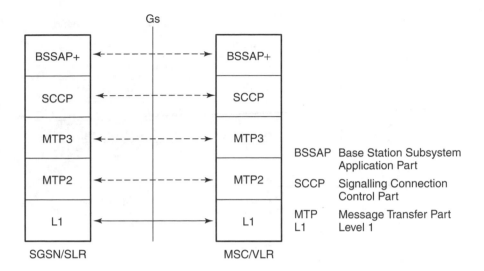

Figure 3.25   The interface Gs

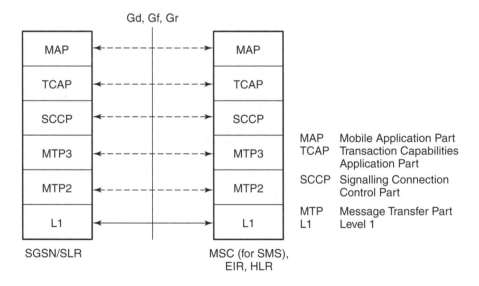

Figure 3.26   The SS7 interfaces Gd, Gf, Gr

The Gd interface between the SGSN and an MSC connected to the SMS Server, is necessary for the transport of SMS if the SMS Centre is only connected to an MSC and not to an SGSN.

The Gr, Gf and Gd interfaces (see Figure 3.26) are based on the MAP (Mobile Application Part) protocol which is the application layer of all signalling interfaces in the GSM Core Network.

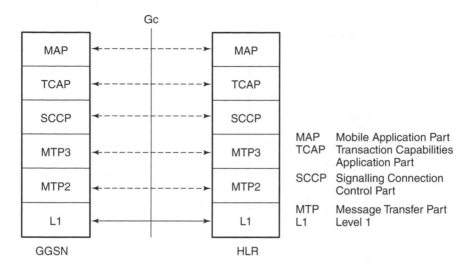

Figure 3.27   The optional direct interface Gc

Below the application layer the Gs, Gr, Gf and Gd interfaces are all based on SCCP (Signalling Connection Control Part), MTP 2 and 3 (Message Transfer Part – levels 2 and 3) and a level 1. Different solutions are possible:

- These signalling interfaces are implemented on a classical Time Division Multiplex TDM network (on the protocol stack 'L1' refers to the TDM network). The information is physically transported on E1 links (2048 kbit/s). MTP 2 and 3 handle the usual tasks of the SS7 levels 2 and 3 – i.e. MTP 2 error correction and MTP 3 the addressing.
- These signalling interfaces are implemented on an ATM network or on an IP network (on the protocol stack 'L1' refers to the ATM or the IP network). The physical realization may differ from one implementation to another. MTP 2 and 3 are modified for these networks but handle the same tasks.

Above MTP 2 and 3 everything is identical in every case – i.e. SCCP does not 'know' which solution is implemented in the lower layers.

3.9.3.2   The Optional Gc Interface

Gc is an optional interface between the GGSN and the HLR. This interface is necessary for incoming (network originated) data applications – e.g. GPRS push services. The Gc interface is based on MAP (Mobile Application Part), therefore the protocol stack is similar to the ones of the Gs, Gr, Gf and Gd interfaces.

There are two alternative ways to implement this interface:

- If an SS7 interface is installed in the GGSN, the MAP protocol can be used between the GGSN and an HLR (see Figure 3.27).
- If an SS7 interface is not installed in the GGSN, any GSN with a SS7 interface installed in the same PLMN as the GGSN, can be used as a GTP-to-MAP protocol converter to allow signalling between the GGSN and an HLR (see Figure 3.28).

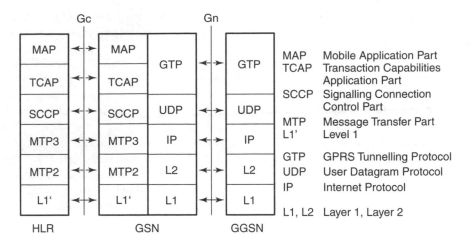

Figure 3.28   The optional interface Gc via a GSN node

3.9.3.3   SS7 Introduction

In this and other texts many references are made to SS7 with regard to signalling. The following section provides a brief introduction to SS7 along with an example of the use of SS7 in GPRS.

The Signalling System No 7 is one of the most important protocols used mainly in core networks of both Integrated Services Digital Networks (ISDN) as well as GSM Public Land Mobile Networks (PLMN). It is a type of out-of-band signalling and a prerequisite for an ISDN network. Since the GPRS network elements communicate with HLR, MSC/VLR, etc. the protocol these elements use has to be deployed on these interfaces. The goal of this introduction is to allow the reader to have a clearer idea of what SS7 is.

SS7 is used to enable network elements to communicate with each other in order to perform such tasks as call setup and database interrogation. The different functions of this protocol are specified in the SS7 model. The lowest three levels form the MTP (Messages Transfer Part) and correspond roughly to OSI Layer 1, 2 and 3. The MTP transfers messages sent by the user parts (level 4 – e.g. ISUP, TUP, SCCP, see exemplary protocol stack in Figure 3.29). Some of the main MTP tasks are:

- secure message transfer
- message routing
- physical transmission of the bits

Above the MTP layer are the so called 'User Parts':

- The Telephone User Part (TUP) and the ISDN User Part (ISUP) are used e.g. to setup, maintain and clear-down telephone calls.
- The Signalling Connection Control Part (SCCP) enables the use of Global Titles for the transfer of signalling information without a relationship to a user channel.

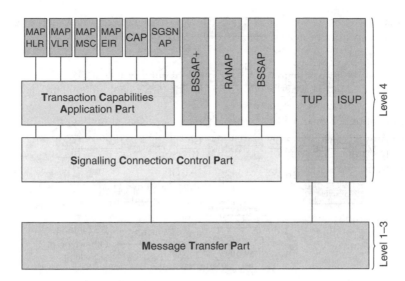

Figure 3.29 SS7 model – protocol stack with MTP and some exemplary user and application parts

Above the SCCP are the so called 'Application Parts' (AP).

- The SGSN AP is used in the SGSN to communicate on the Gr interface with the MAP HLR in the HLR. It allows database enquiries, e.g. for mobility management.
- The BSSAP+ deployed on the Gs interface supports those MS that use both GSM and GPRS. Through this application part, combined procedures are possible, e.g. paging for GSM via the GPRS network.

Figure 3.30 gives an overview of where SS7 is deployed in a GSM/GPRS network.

To equip the reader with a clearer picture of a typical signalling message exchange, an example of an SS7 signalling procedure is shown in the case of the GPRS attachment procedure (see Figure 3.31).

The SGSN contacts the HLR for a GPRS location update. In the SGSN AP message its SGSN address and the IMSI are sent. Thereafter the database entry in the old SGSN is cancelled. The HLR enters a the subscriber data in the new SGSN and acknowledges the GPRS location update.

Since SS7 is only deployed on the Gr interface, only a part of the whole procedure is performed using SS7. The messages and their parameters vary depending on the user or application part exchanging information.

3.10 GPRS Protocol Planes

3.10.1 GPRS Transmission Plane

Figure 3.32 gives a complete overview of the GPRS transmission plane, which we have so far examined only detail by detail. Apparently the application layer data is transmitted

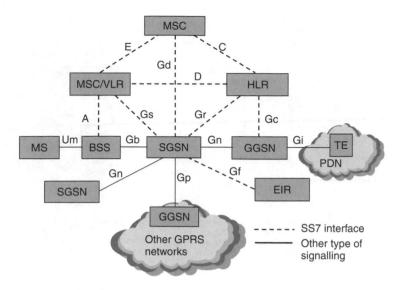

Figure 3.30   SS7 interfaces in the GSM/GPRS network

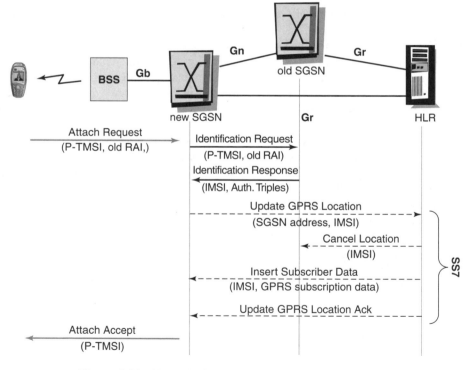

Figure 3.31   Example for SS7 signalling on the Gr interface

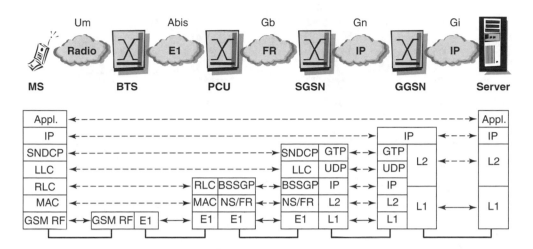

Figure 3.32   The GPRS transmission plane

transparently between the MS and the PDN (Packet Data Network). For this application the MS receives a temporary IP address from the GGSN or perhaps directly from the PDN. The Core Network is also based on IP (Gn interface). On the radio side, the compression and segmentation of the IP packets is done by the SNDCP layer between the MS and the SGSN. Also, a logical link is defined between them by LLC. The Gb interface is based on Frame Relay and the radio interface has the two layers RLC and MAC for segmenting the information into radio blocks and for the allocation (and sharing) of resources.

3.10.2   GPRS Signalling Planes

3.10.2.1   GPRS Signalling Plane MS-SGSN

The GMM and SM messages are exchanged between the MS and the SGSN (see Figure 3.33). These signalling messages are based on LLC frames – i.e. the Gb and Um interfaces are implemented in the same way as in the GPRS transmission plane.

3.10.2.2   The Other Signalling Interfaces

The signalling interfaces inside the Core Network – i.e. the Gc, Gf, Gr and Gs interfaces – are all based on SCCP (Signalling Connection Control Part), MTP 2 and 3 (Message Transfer Part, levels 2 and 3) and a level 1.

The messages themselves are BSSAP+ messages for the Gs interface and MAP messages for the other three interfaces.

3.10.3   GPRS SMS Plane

The GPRS SMS Plane between the MS and the SGSN is similar to the GPRS Signalling Plane between these two elements, except that it is used for the transport of the Short Message Service (the SMS layer is indicated with a specific SAPI in the LLC layer).

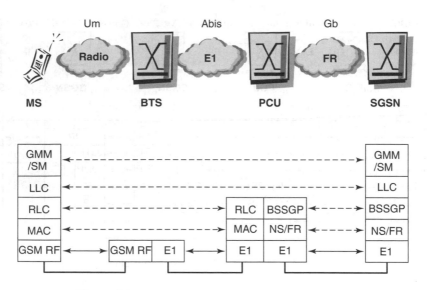

Figure 3.33   The GPRS signalling plane MS – SGSN

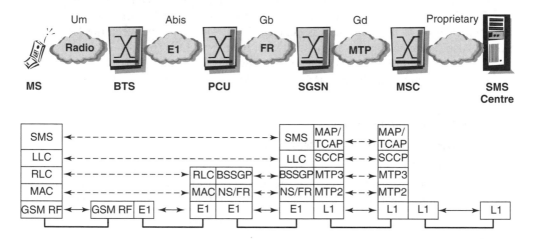

Figure 3.34   The GPRS SMS plane

After the Gb interface, the SGSN converts the SMS into a MAP message for the Gd interface which is implemented like the signalling interfaces of the Core Network. The SMS is sent to an MSC which is connected to the SMS Server. The interface between this MSC and the SMS Server is not specified – i.e. depends on the proprietary solution. Figure 3.34 gives an overview of all net elements and protocols involved.

4

GPRS Procedures

4.1 GPRS Mobility Management Procedures

In this section, the different procedures defined in the GPRS mobility management (GMM) are presented. The GMM protocol is transparent between the MS and the SGSN. This protocol not only includes all the procedures necessary for handling the mobility of the subscriber (e.g. routing area updates), but also the security procedures (e.g. the authentication procedure).

4.1.1 Security Procedures

The SIM card and the authentication centre both contain important information for the different security procedures:

- An individual identification key, Ki, which serves as a password for the subscriber.
- The ciphering key, GPRS Kc, which is calculated for every single connection to avoid eavesdropping.
- Two algorithms to calculate the respective keys, i.e. A3 for the authentication key and A8 for the ciphering key.

Two other MS identifiers are stored in the SIM card and in the SGSN:

- The International mobile subscriber identity (IMSI), which serves as a fixed user name to identify the user towards the network. It consists of the mobile country code (MCC), the mobile network code (MNC) and the mobile station identity number (MSIN). The IMSI is also stored in the HLR and in the AC.
- The Packet Temporary Mobile Subscriber Identity (P-TMSI), which serves as a temporary user name to identify the user towards the network. This temporary name is allocated by the SGSN, which may also regularly reallocate a new P-TMSI to the MS.

The algorithm GPRS A5 is the algorithm that actually ciphers the information. It is implemented in the mobile equipment (ME) and in the SGSN. Therefore the information is ciphered between the MS and the SGSN and not only over the air interface, as is the

GPRS Networks. G. Sanders, L. Thorens, M. Reisky, O. Rulik, S. Deylitz
© 2003 John Wiley & Sons, Ltd. ISBN: 0-470-85317-4

case with circuit switched GSM calls. The ciphering is a task performed by the LLC protocol which is transported transparently between the MS and the SGSN, i.e. LLC information is not decoded in the base station.

4.1.1.1   Authentication

As soon as the MS makes contacts with the network, it has to be authenticated before it is allowed to have access. For example, authentication is performed at the start of:

- a routing area update,
- a GPRS attach or detach,
- a GPRS packet transfer.

The goal of the authentication procedure is to certify that the subscriber who is trying to initiate a connection with the network has an authentic SIM card with a valid Ki. This must be verified *without* sending the Ki over the radio interface. The authentication procedure is initiated and controlled by the SGSN. The two other important elements involved are of course the MS and the AC.

Before the MS may exchange any data with the network, the SGSN initiates an authentication for this MS. First the SGSN sends a message containing the IMSI of the subscriber to the AC and requests some triples. Using the IMSI, the AC finds the associated Ki and so can calculate the necessary keys and send them to the VLR. A triple is composed of three keys called RAND, SRES and Kc, which are explained below:

- RAND is a random number used to ensure that triples are never the same;
- SRES (signed response) can be thought of as a digital signature for the MS;
- Kc is the ciphering key used to encrypt the data sent between MS and SGSN.

The digital signature (SRES) is produced by an algorithm called 'A3'. The ciphering key (Kc) is produced by another algorithm called 'A8'. The two algorithms A3 and A8 both have the same inputs: Ki and RAND, i.e. one input which is stored in the AC and is always the same (Ki), and one input which is always different (RAND).

To authenticate the subscriber, the SGSN sends RAND to the MS. The SIM card also contains the algorithms A3 and A8, and the individual authentication key, Ki. When the SGSN sends RAND to the MS, the SIM card then has everything it needs to calculate the keys SRES and Kc. As the keys are based on the same inputs used by the AC (Ki and RAND), the values calculated by the SIM should be the same. The SIM card then sends the SRES back to the SGSN which compares the SRES from the AC, with the SRES from the SIM card. If these two SRES are the same, the authentication is successful and the SGSN sends a confirmation to the MS, if not the SGSN sends a message 'authentication not valid' to the MS (Figure 4.1).

The whole idea of this procedure is to check the SIM card of the MS – i.e. the Ki – without sending this Ki over the radio interface. A common cause for concern is that all these messages over the radio interface are sent unencrypted because the ciphering starts after the authentication. Security conscious users worry that if someone manages to intercept RAND and SRES as they are transmitted over the radio interface, and if this person knows the algorithm A3, it may be possible to reverse the calculation to derive Ki.

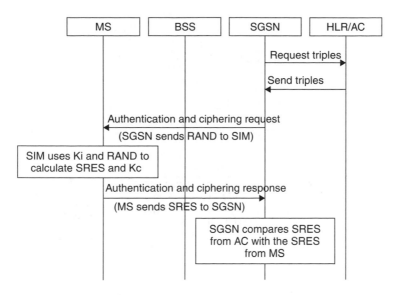

Figure 4.1 The authentication procedure

In fact the algorithms used in GSM are designed to make it extremely difficult to calculate the input Ki (128 bits) from the output SRES. Such inverse problems may require a lot of processing time on a good computer to find the solution.

A final point on the subject of authentication concerns the excessive signalling between the HLR/AC and the SGSN that would result if every SGSN requested a single triple from the AC for every authentication. The AC is usually configured to send several triples in response to a request. These are then stored and used by the SGSN, which will then only request more when it needs them. An AC might (typically) be configured to send five triples at a time to the SGSN which can then carry out five authentications for the subscriber before it needs to make another request.

4.1.1.2 Ciphering

In order to avoid confusion, the start of ciphered transmission between MS and the SGSN has been predefined. The MS starts ciphering its transmissions after sending the authentication response message, and the SGSN starts ciphering when a valid authentication response message is received from the MS.

Ciphered transmission is enabled through the use of the ciphering key, GPRS Kc, and the ciphering algorithm, GPRS A5. The SGSN receives GPRS Kc as part of the triple, while the MS calculates GPRS Kc after receiving RAND.

There are notable differences between ciphering in GSM and GPRS. In GSM, ciphering is performed between MS and BTS and uses one of three versions of A5 (A5-0, A5-1 or A5-2), depending on the level of ciphering permitted. In GPRS, ciphering is performed between the MS and the SGSN and can use a new version of A5 developed especially for packet transmission (A5-3).

GPRS Kc is one input parameter of GPRS A5 and the output is a ciphering bit sequence (Ciph-S), which then ciphers a bit sequence on the transmission side and deciphers this

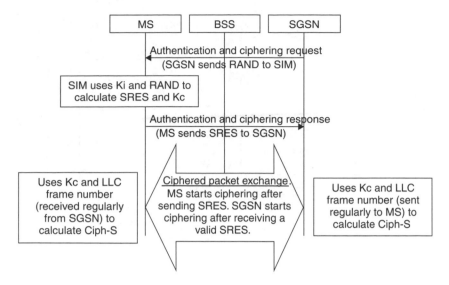

Figure 4.2   The ciphering procedure

ciphered bit sequence on the receiving side. If GPRS Kc was the only one input parameter, the ciphering bit sequence would be the same for the whole GPRS session, so there is another input parameter which depends on the LLC frame number. Therefore each LLC frame is ciphered with a different Ciph-S. Of course, the SGSN must regularly send the LLC frame number to the MS, such that the MS and SGSN stay in sync.

The ciphering/deciphering operations themselves (Figure 4.2) are performed using logical XOR summation, which is described in the following example. To keep things simple, 8-bit sequences are used rather than real LLC frames.

– the bit sequence to be transmitted is 01110010
– the ciphering sequence (Ciph-S) is 11100110.

The XOR summation is the bit wise summation of these two sequences using the following rules:

$$0 + 0 = 0$$

$$0 + 1 = 1$$

$$1 + 0 = 1$$

$$1 + 1 = 0$$

Which means the Ciph-S and the bits to be sent can be combined in the following way:

bits to be ciphered 01110010

Ciph-S 11100110

XOR summation/result 10010100

The resulting bit sequence (10010100) is sent from the MS to the SGSN (or vice versa) where it must be deciphered. This happens as follows:

bits to be deciphered 10010100

Ciph-S 11100110

XOR summation/result 01110010

In the example, the output from the ciphering process is transmitted and deciphered on the receiving side using the same ciphering sequence. The result is 01110010, which is exactly the original unencrypted bit sequence, thus demonstrating the simple yet effective process used for ciphering. As mentioned previously, the ciphering is performed in the LLC layer which means the actual Ciph-S must be the same length as the LLC frame being ciphered. In reality, the length of the LLC frames is variable and may be up to 1523 octets long.

The fact that the ciphering sequence is based on the LLC frame number and that GPRS Kc is never sent over the air interface ensures that it would be extremely difficult to eavesdrop on the connection over the radio interface.

4.1.1.3 P-TMSI Reallocation

The packet temporary mobile subscriber identity (P-TMSI) serves as a temporary user name to identify the user towards the network. The P-TMSI takes the place of the permanent user identity, the IMSI, which should be sent over the air interface as infrequently as possible. The unencrypted P-TMSI is sent to the SGSN at the start of any procedure between the MS and the network. We will see later (in the section on routing area update and GPRS attach) that sometimes the MS has to send its IMSI because it has no P-TMSI or the P-TMSI is not valid.

This temporary identity is allocated by the SGSN after the authentication procedure and the start of ciphering, i.e. after every authentication the MS receives a new P-TMSI. Also, when the MS moves to a new SGSN area, the new SGSN allocates a new P-TMSI and the P-TMSI of the first SGSN is then free for the next subscriber in its area. Each SGSN has such a pool of P-TMSIs. If the MS stays with the same SGSN for a long time, this SGSN may allocate a new P-TMSI to the MS.

4.1.1.4 Identity Check Procedure

As explained in Chapter 2, the mobile network may implement an equipment identity register (EIR) which checks the status of the mobile equipment (ME) as it could be stolen.

If an EIR is implemented, the identity check procedure may be done after the authentication and the start of the ciphering. The SGSN sends an identity request to the MS which responds with the international mobile equipment identity (IMEI). The SGSN then forwards this IMEI to the EIR which checks this IMEI against the different lists and especially checks that this IMEI is not on the black list. Then the EIR sends a message back to the SGSN that indicates whether the MS was found on to the white, grey or black list.

4.1.2   Location Management Procedures

The location management procedures describe the actions necessary when the MS is GPRS attached (the MS is either in STANDBY or in READY state) and moves from cell to cell. For the GPRS attach procedure, see Section 4.1.3.1.

Each cell in the network provides general information on its broadcast control channel (BCCH). The information important for the location management procedures are cell global identity (CGI), the routing area identity (RAI) and the location area identity (LAI).

When the MS moves to a new cell, which is possibly part of a new routing area or even a new location area, the following three scenarios are possible:

- the MS initiates a cell update,
- the MS initiates a routing area update,
- the MS initiates a combined routing area and location area update.

If there is no Gs interface between the current SGSN and the current MSC/VLR, and if the MS determines that it shall perform both a location area and a routing area update, the MS shall perform the location area update first, i.e. the mobile decodes the BCCH of the new cell and is informed that: network operation mode 2 or 3 (NOM-II/III) is in use (no Gs), LAI has changed and RAI has changed. It should be noted that if NOM-I is in use, the updates are handled via CMM in the SGSN.

4.1.2.1   Cell Update

A cell update takes place when the MS enters a new cell inside the current routing area and the MS is in READY state (in STANDBY state, no cell updates are performed, only routing area updates). If the routing area has changed, a routing area update is executed instead of a cell update.

The MS performs the cell update procedure by sending a message containing the P-TMSI to the SGSN. In the direction towards the SGSN, the PCU shall add the cell global identity CGI. The SGSN records this MS's change of cell, and further traffic directed towards the MS is conveyed over the new cell.

4.1.2.2   Routing Area Update

In this section it is assumed that either (a) the MS only moves to a new routing area and not to a new location area (in this case it is an intra SGSN routing area update) or (b) no Gs interface is implemented, so if the MS determines that it shall perform both a location area and a routing area update, the MS shall perform the location area update first. Combined routing area and location area updates with the Gs interface are described in Section 4.1.2.3.

A routing area update takes place when a GPRS attached MS detects that it has entered a new routing area (RA). There are two different routing area updates: either it is an intra-SGSN routing area update, i.e. both routing areas are controlled by the same SGSN, or it is an inter-SGSN routing area update, i.e. each routing area is controlled by a different SGSN.

Intra-SGSN Routing Area Update

In this case, the MS moves to a new routing area that is controlled by the same SGSN, so the SGSN only has to store the new routing area identity RAI, and no other element (HLR or GGSN) needs to receive information about this update. The procedure, which is illustrated in Figure 4.3, is described in more details in the following sequence:

(1) The MS sends a routing area update request with its P-TMSI and the old routing area identity (RAI) to the SGSN. The PCU shall add the new cell global identity (CGI) including the new routing area code (RAC) of the cell where the message was received before passing the message to the SGSN.

(2) Security functions may be executed: authentication, start of ciphering and optional equipment identity check.

(3) The SGSN validates the MS's presence in the new RA. The SGSN detects that an intra-SGSN routing area update must be performed by checking that it is also responsible for the old RA, i.e. the SGSN covers both the old RA and the new RA. In this case, the SGSN has all the necessary information about the MS and there is no need to inform the GGSN or the HLR about the new MS location. A routing area update acceptance (possibly with the allocation of a new P-TMSI) is returned to the MS.

(4) If the P-TMSI was reallocated, the MS acknowledges the new P-TMSI by returning a routing area update complete message to the SGSN.

Inter-SGSN Routing Area Update

In this case, the MS moves to a new routing area, which is controlled by a new SGSN; therefore the new SGSN has to store the details of the new subscriber along with the routing area identity. Further, the old SGSN must forward any buffered packets and any packets that arrive during the changeover to the new SGSN. Additionally, the HLR, and perhaps one or more GGSNs (in case of active PDP contexts), must also be notified of the SGSN change. The following sequence, illustrated in Figure 4.4, describes the procedure in more detail:

(1) The MS sends a routing area update request with its P-TMSI and the old RAI to the new SGSN. The PCU shall add the CGI including the RAC of the cell where the message was received before passing the message to the SGSN.

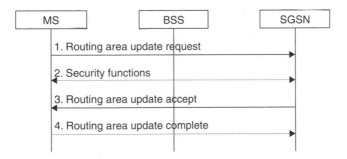

Figure 4.3   Intra-SGSN routing area update

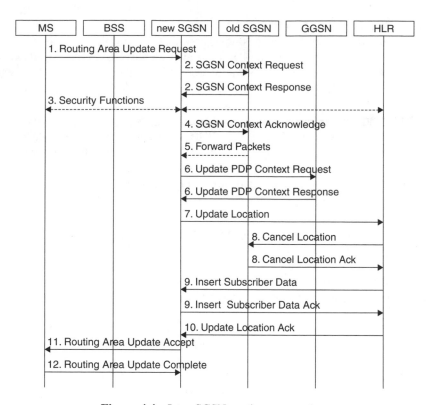

Figure 4.4   Inter-SGSN routing area update

(2) The domain name server DNS of the mobile network contains a central database with all routing areas of the mobile network coupled with their corresponding SGSN. Therefore, thanks to the DNS, the new SGSN is able to find the old SGSN with the old RAI. The new SGSN sends an SGSN context request, including its own IP address, the P-TMSI and the old RAI, to the old SGSN to get the MM and PDP contexts for the MS. The old SGSN stores the IP address of the new SGSN such that it can forward data packets to the new SGSN. The old SGSN recognizes the P-TMSI of the MS and sends a message back containing the IMSI of the MS, the possible remaining triples (for future authentication and ciphering procedures) and the situation of all active PDP contexts (the number of sent and received messages on the SNDCP layer for the Um+Gb interface and on the GTP layer for the Gn interface) to the new SGSN. The old SGSN starts a timer (explained in Step 5) and stops the transmission of SNDCP packets to the MS.

(3) Security functions managed by the new SGSN may be executed: authentication, start of ciphering and optional identity check.

(4) The new SGSN sends an SGSN context acknowledge message to the old SGSN. This informs the old SGSN that the new SGSN is ready to receive data packets belonging to the activated PDP contexts.

(5) The old SGSN duplicates the buffered SNDCP packets and starts tunnelling them to the new SGSN. Additional SNDCP packets received from the GGSN before

the timer described in Step 2 expires are also duplicated and tunnelled to the new SGSN. SNDCP packets already sent to the MS in acknowledged mode and not yet acknowledged by the MS are tunnelled to the new SGSN as well. No SNDCP packets are forwarded to the new SGSN after expiry of the timer described in Step 2.

(6) The new SGSN sends an update PDP context request including its own IP address, the tunnel identifier (TID) and the negotiated quality of service to the concerned GGSNs. The GGSNs update their PDP context fields and each returns an update PDP context response (with the TID). For more detailed information about the PDP contexts, see Section 4.2 of this Chapter

(7) The new SGSN informs the HLR of the change of SGSN by sending an update location message to the HLR which includes the SGSN's SS7 identifier, the SGSN's IP address and the IMSI of the MS.

(8) The HLR sends a cancel location message to the old SGSN, along with the IMSI. The old SGSN removes the MM and PDP contexts of the MS only when the timer described in Step 2 expires. This allows the old SGSN to complete the forwarding of SNDCP packets. The old SGSN sends a cancel location acknowledge message with the IMSI to the HLR.

(9) The HLR sends an insert subscriber data message with the IMSI and the GPRS subscription data to the new SGSN. The SGSN constructs an MM context for the MS and returns an insert subscriber data ack message with the IMSI to the HLR.

(10) The HLR acknowledges the update location by sending an update location acknowledgement message with the IMSI to the new SGSN.

(11) A logical link is established between the new SGSN and the MS. The new SGSN sends a routing area update accept message, which includes the new P-TMSI and the number of UL received SNDCP packets (confirming all mobile originated SNDCP packets successfully transferred before the start of the update procedure) to the MS.

(12) The MS acknowledges the new P-TMSI by returning a routing area update complete message which includes the number of DL received SNDCP packets (confirming all mobile terminated SNDCP packets successfully transferred before the start of the update procedure) to the SGSN.

Inter-Network Routing Area Update
In this case the MS moves to a new mobile network, for example when crossing the border between two countries (if the two networks concerned have a roaming agreement with each other). The update procedure is almost the same as that described in the previous section. The only difference is that in Step 2, the domain name server (DNS) of the new network contacts the DNS of the old network to get the IP address of the old SGSN based on the old routing area identity.

4.1.2.3 Combined Routing Area/Location Area Update

In this section it is assumed that the Gs interface between the SGSN and the MSC/VLR is implemented. In which case, the MS may request a combined routing area and location area update when it moves to a new location area. It should be noted that since routing areas are always smaller than or equal to location areas, and that a routing area must lie fully within one location area, a location area change will always coincide with a routing area change. The MS only contacts the SGSN, which saves the new routing area identity and contacts the VLR (via the Gs interface), which then stores the new location area identity (LAI). Thanks to the implementation of the Gs interface, only one update request over the radio interface is necessary.

Combined Intra-SGSN Routing Area/Intra-MSC Location Area Update

In this case the MS moves to a new location area which is also controlled by the current MSC/VLR, and to a new routing area also controlled by the current SGSN. Therefore, the VLR has to store the new location area identity and the SGSN the new routing area identity, but no other element (HLR or GGSN) needs to receive information about this update. The procedure is illustrated in Figure 4.5 and described in more detail in the following sequence:

(1) The MS sends a combined routing area/location area update request with its P-TMSI and the old routing area identity (RAI) containing the old LAI, to the SGSN. The PCU shall add the new cell global identity (CGI) including the new routing area code (RAC) and the new location area code (LAC) of the cell where the message was received before passing the message to the SGSN.
(2) Security functions may be executed: authentication, start of ciphering and optional equipment identity check.
(3) The SGSN sends a location update request containing the new LAI, the IMSI and the SGSN SS7 identifier to the VLR. The VLR SS7 identifier is translated from the new RAI via a table in the SGSN. The SGSN SS7 identifier is stored in the VLR.
(4) The new VLR optionally allocates a new VLR TMSI and responds with a location update accept message that includes this optional VLR TMSI to the SGSN. The VLR TMSI reallocation is optional because the VLR has not changed.
(5) The SGSN validates the MS's presence in the new RA. A routing area update accept message (possibly with the allocation of a new P-TMSI and of a new TMSI from the VLR) is returned to the MS.

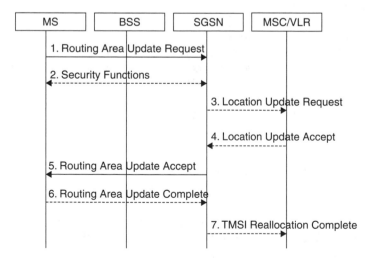

Figure 4.5 Combined intra-SGSN routing area/intra-MSC location area update

(6) If the P-TMSI and/or VLR TMSI was reallocated, the MS acknowledges the new P-TMSI and/or the new VLR TMSI by returning a routing area update complete message to the SGSN.

(7) The SGSN sends a TMSI reallocation complete message to the VLR if the VLR TMSI is confirmed by the MS.

Combined Intra-SGSN Routing Area/Inter-MSC Location Area Update

In this case the MS moves to a new location area that is controlled by a new MSC/VLR and to a new routing area which is controlled by the same SGSN. This case is especially likely at the introduction of GPRS, because the network will almost certainly have far fewer SGSNs than MSCs. Therefore the new VLR has to store the new location area identity (LAI) and the SGSN must store the new routing area identity (RAI); additionally the HLR must store the identity of the new VLR. The procedure is illustrated in Figure 4.6. and described in more detail in the following sequence:

(1) The MS sends a combined routing area/location area update request with its P-TMSI and the old routing area identity (RAI) containing the old LAI, to the SGSN. The PCU will add the new cell global identity (CGI) including the new routing area code (RAC) and the new location area code (LAC) of the cell where the message was received before passing the message to the SGSN.

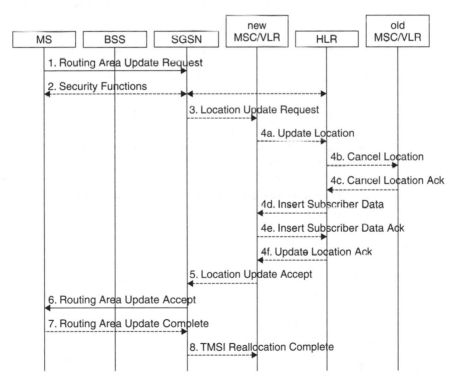

Figure 4.6 Combined intra-SGSN routing area/inter-MSC location area update

(2) Security functions may be executed: authentication, start of ciphering and optional equipment identity check.

(3) The SGSN sends a location update request containing the new LAI, the IMSI and the SGSN SS7 identifier to the new VLR. The new VLR SS7 identifier is translated from the new RAI via a table in the SGSN. The new VLR stores the SGSN SS7 Identifier.

(4) The GSM location update is performed as follows:

   (a) The new VLR sends an update location request, with its own VLR SS7 identifier and the IMSI of the MS, to the HLR.

   (b) The HLR cancels the data in the old VLR by sending cancel location message with the IMSI to the old VLR.

   (c) The old VLR acknowledges with a cancel location acknowledge message containing the IMSI of the MS being cancelled.

   (d) The HLR sends an insert subscriber data message, including the IMSI and the GSM subscriber data of the MS, to the new VLR.

   (e) The new VLR acknowledges with an insert subscriber data acknowledge message containing the IMSI of the MS.

   (f) The HLR responds with an update location acknowledge message, containing the IMSI of the MS, to the new VLR.

(5) The new VLR allocates a new VLR TMSI and responds with a location update accept message, including this new VLR TMSI, to the SGSN.

(6) The SGSN validates the MS's presence in the new RA. A routing area update Accept message (possibly with the allocation of a new P-TMSI and with the new TMSI from the VLR) is returned to the MS.

(7) The MS acknowledges the new VLR TMSI (and possibly the new P-TMSI if the SGSN allocated a new one) by returning a routing area update complete message to the SGSN.

(8) The SGSN sends a TMSI reallocation complete message to the VLR.

Combined Inter-SGSN Routing Area/Inter-MSC Location Area Update

In this case, the MS moves to a new location area that is controlled by a new MSC/VLR, and to a new routing area controlled by a new SGSN. Therefore, the new VLR has to store the new location area identity (LAI) and the new SGSN the new routing area identity (RAI); also the HLR and maybe some GGSNs (in case of active PDP contexts) must be notified about the new SGSN and the new VLR. The procedure is illustrated in Figure 4.7. and described in more detail in the following sequence:

(1) The MS sends a combined routing area/location area update request with its P-TMSI and the old routing area identity (RAI) containing the old LAI, to the SGSN. The PCU shall add the new cell global identity (CGI) including the new routing area code (RAC) and the new location area code (LAC) of the cell where the message was received before passing the message to the new SGSN.

(2) The domain name server (DNS) of the mobile network contains a central database with all routing areas of the mobile network coupled with their corresponding SGSN. Therefore, thanks to the DNS, the new SGSN is able to find the old SGSN using the old RAI. The new SGSN sends an SGSN context request including its own IP address, the P-TMSI and the old RAI to the old SGSN to get the MM and PDP

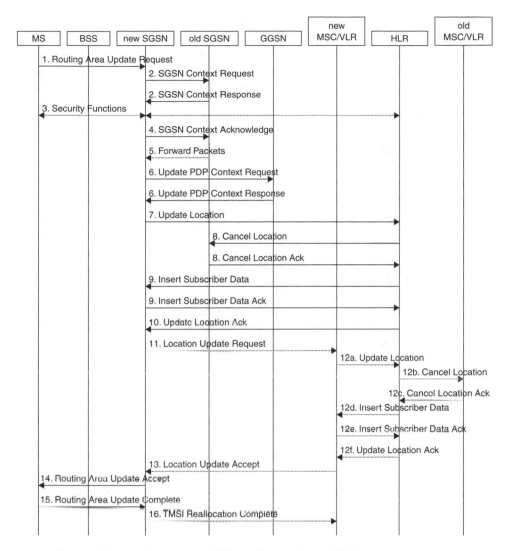

Figure 4.7 Combined inter-SGSN routing area/inter-MSC location area update

contexts for the MS. The old SGSN stores the IP address of the new SGSN such that it can forward data packets to the new SGSN. The old SGSN recognizes the P-TMSI of the MS and sends a message back containing the IMSI of the MS, any remaining unused triples (for future authentication and ciphering procedures) and the details of all active PDP contexts (the number of sent and received messages on the SNDCP layer for the Um+Gb interface and on the GTP layer for the Gn interface) to the new SGSN. The old SGSN starts a timer (explained in point 5)and stops the transmission of SNDCP packets to the MS.

(3) Security functions managed by the new SGSN may be executed: authentication, start of ciphering and optional equipment identity check.

(4) The new SGSN sends an SGSN context acknowledge message to the old SGSN. This informs the old SGSN that the new SGSN is ready to receive data packets belonging to the activated PDP contexts.

(5) The old SGSN duplicates the buffered SNDCP packets and starts tunnelling them to the new SGSN. Additional SNDCP packets received from the GGSN before the timer described in step 2 expires, are also duplicated and tunnelled to the new SGSN. SNDCP packets that were already sent to the MS in acknowledged mode and which have not yet been acknowledged by the MS are tunnelled to the new SGSN as well. No SNDCP packets are forwarded to the new SGSN after expiry of the timer described in step 2.

(6) The new SGSN sends an update PDP context request, including its own IP address, the tunnel identifier (TID) and the negotiated quality of service to the concerned GGSNs. The GGSNs update their PDP context fields and each returns an update PDP context response (with the TID). For more detailed information about the PDP contexts, see section 4.2 of this chapter.

(7) The new SGSN informs the HLR of the change of SGSN by sending an update location message, which includes the SGSN SS7 identifier, the SGSN IP address and the IMSI of the MS, to the HLR.

(8) The HLR sends a cancel location message, including the IMSI, to the old SGSN. The old SGSN removes the MM and PDP contexts of the MS only when the timer described in Step 2 expires. This allows the old SGSN to complete the forwarding of SNDCP packets. The old SGSN sends a cancel location acknowledge message, with the IMSI, to the HLR.

(9) The HLR sends an insert subscriber data message, with the IMSI and the GPRS subscription data, to the new SGSN. The SGSN constructs an MM context for the MS and returns an insert subscriber data acknowledge message with the IMSI to the HLR.

(10) The HLR acknowledges the location update by sending an update location acknowledge message, with the IMSI, to the new SGSN.

(11) The new SGSN sends a location update request, containing the new LAI, the IMSI and the SGSN SS7 Identifier, to the new VLR. The new VLR SS7 identifier is translated from the new RAI via a table in the new SGSN. The new VLR stores the SGSN SS7 Identifier.

(12) The GSM location update is performed as follows:

    (a) The new VLR sends an update location request, with its own VLR SS7 identifier and the IMSI of the MS, to the HLR.

    (b) The HLR cancels the data in the old VLR by sending a cancel location message, with the IMSI, to the old VLR.

    (c) The old VLR acknowledges with a cancel location acknowledge message containing the IMSI.

    (d) The HLR sends an insert subscriber data message, including the IMSI and the GSM subscriber data of the MS, to the new VLR.

    (e) The new VLR acknowledges with an insert subscriber data acknowledge message containing the IMSI.

    (f) The HLR responds with an update location acknowledge message, containing the IMSI, to the ncw VLR.

(13) The new VLR allocates a new VLR TMSI and responds with a location update accept message, including this new VLR TMSI, to the new SGSN.

(14) The new SGSN validates the MS's presence in the new RA and allocates a new P-TMSI. A routing area update accept message, with the new P-TMSI and the new VLR TMSI, is returned to the MS.

(15) The MS acknowledges the new P-TMSI and the new VLR TMSI by returning a routing area update complete message to the new SGSN.

(16) The new SGSN sends a TMSI reallocation complete message to the new VLR.

4.1.3 GPRS Attach and Detach Procedures

In the GPRS attach procedure, the MS changes its state from IDLE to READY. This may be done when the user switches on his mobile device or when the user explicitly activates GPRS on his already IMSI attached end device.

In the GPRS detach procedure, the MS changes its state from READY or STANDBY to IDLE. This may be done when the user switches off his mobile device or when the user explicitly deactivates GPRS on his IMSI attached mobile device.

4.1.3.1 GPRS Attach

The GPRS attach procedure is very similar to the routing area update procedures described previously. Therefore, only the differences (rather than the full descriptions of the different cases) are presented here.

There are two differences:

- No messages about PDP contexts are exchanged (for example between two SGSNs as in the inter SGSN routing area update procedure), because all PDP contexts are inactive as the MS was GPRS detached.
- If the SIM card is new, in the first step, i.e. the attach request, the MS has to send this request with the IMSI and not with any P-TMSI, because the MS has no P-TMSI.

The GPRS attach procedure may also be combined with the IMSI attach procedure if the Gs interface is implemented between the MSC/VLR and the SGSN. If the Gs interface is not implemented and if the MS wants to perform both a GPRS and an IMSI attach, the MS shall perform the IMSI attach first.

4.1.3.2 GPRS Detach

The (explicit) GPRS detach function allows the MS to inform the network that it wants to make a GPRS detach or a combined GPRS/IMSI detach. An implicit detach is also possible after the so called 'mobile reachable' timer has expired, or after an irrecoverable radio error causes disconnection of the logical link. In each case, the network detaches the MS without notifying it (because in each case the MS in no longer reachable).

In the request message of a GPRS detach procedure initiated by the MS, there is an indication to inform the SGSN whether the detach is due to the device being switched off or not. This is done because, if the device is being switched off, the SGSN does not send any detach accept message to the MS (because it would not be received).

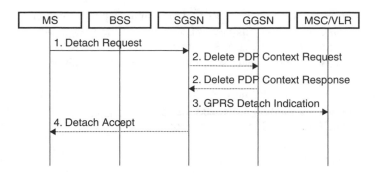

Figure 4.8 GPRS detach initiated by the MS (without IMSI detach request and with the Gs interface)

The GPRS detach procedure may also be combined with the IMSI detach procedure if the Gs interface is implemented between the MSC/VLR and the SGSN. If the Gs interface is not implemented and if the MS wants to perform both a GPRS and an IMSI detach, the MS performs the GPRS detach first.

The GPRS detach procedure initiated by the MS (without an IMSI detach request and with the implementation of the Gs interface) is illustrated in Figure 4.8. and described in the following sequence:

(1) The MS sends a GPRS detach request (without an IMSI detach request) including the P-TMSI to the SGSN.
(2) The active PDP contexts in the GGSNs regarding this particular MS are deactivated by the SGSN sending a delete PDP context request with the tunnel identifiers (TID) to the GGSNs. The GGSNs acknowledge with a delete PDP context response containing the corresponding TID.
(3) The MS wants to remain IMSI attached, i.e. still reachable for voice and SMS, and so is only performing a GPRS detach. Therefore the SGSN sends a GPRS detach indication containing the IMSI to the VLR. The VLR removes the association with the SGSN and handles paging and location updates without going via the SGSN.
(4) The detach is not due to the mobile device being switched off, hence the SGSN sends a detach accept to the MS.

The combined GPRS/IMSI detach procedure initiated by the MS is illustrated in Figure 4.9 and described in the following sequence:

(1) The MS sends a combined GPRS/IMSI detach request to the SGSN which includes the P-TMSI and an indication that the mobile is being switched off.
(2) The active PDP contexts in the GGSNs regarding this particular MS are deactivated by the SGSN sending a delete PDP context request with the tunnel identifiers (TID) to the GGSNs. The GGSNs acknowledge with a delete PDP context response containing the corresponding TID.
(3) The SGSN sends an IMSI detach indication, with the IMSI, to the VLR.

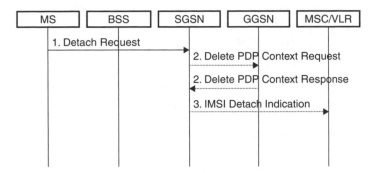

Figure 4.9   Combined GPRS/IMSI detach initiated by the MS

4.2   Session Management Procedures

The PDP (packet data protocol) context of an MS describes the data connection between the MS and an external data network. The PDP context is described in the MS, in the SGSN and in the GGSN, and is either in the inactive or in the active state. A MS may have one or several PDP contexts.

A PDP context may be activated or deactivated at any time if the MS is in STANDBY or in READY state. These procedures are only meaningful at the core network level and in the MS, and do not directly involve the BSS, i.e. the 'session management' layer is transparent between the MS and the SGSN.

If the MS is in IDLE state, all PDP contexts are deactivated and can not be activated before the MS changes its mobility management (MM) state, i.e. a GPRS attach must be performed so the MS goes from IDLE to the READY state.

4.2.1   PDP Context Activation

For the activation of a PDP context, an IP address for the MS is necessary. Two different allocations are possible:

- static allocation: the MS has a permanent IP address – more realistic with IP version 6 (IPv6) as not enough addresses are available with IP version 4 (see Chapter 9 for more details of IP versions);
- dynamic allocation: the corresponding GGSN assigns a temporary IP address to the MS – this is the system used for GPRS with IP version 4 (IPv4).

A PDP context may only be activated by the MS. For mobile terminating packet transfers, e.g. from a news service to which the user subscribes, the GGSN must initiate a paging to the MS that requests PDP context activation in order to receive the packets, that is the GGSN can only request PDP context activation, it can not initiate one. It is important to note here that a PDP context activation request from the GGSN is only possible if the GGSN has static information about the corresponding MS, i.e. if the MS has a static IP address stored in the HLR. The reason for this is that in order for IP packets to be sent to a certain MS, the sender must have an IP address for the MS. Until IPv6 is implemented, an MS will receive a temporary IPv4 address when it activates a PDP context which means

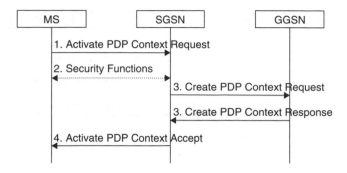

Figure 4.10   PDP context activation initiated by the MS

that mobile terminating packet transfers would be difficult to implement under IPv4 as the service provider would have no direct means of addressing the user's MS.

In the following sequence, illustrated in Figure 4.10, the PDP context activation initiated by the MS is described.

(1) The MS sends an activate PDP context request message containing the following information:
  – Network service access point identifier (NSAPI): this identifier defines the individual application on the SNDCP layer, i.e. each application has a unique NSAPI.
  – The Transaction identifier (TI): this identifier defines the individual sessions on the SM layer.
  – The PDP type: identifies the protocol used transparently between the MS and the GGSN, usually IP (version 4 or version 6) but X.25 is also possible.
  – The PDP address: this field either contains the static IP address of the MS or is empty to signal that the MS needs to be given a dynamic (temporary) IP address.
  – The access point name (APN): this is a logical name identifying the external data network. The APN refers to the interface (Gi) that connects to a certain network.
  – The quality of service (QoS) requested from the SGSN for the application.
(2) Security functions may be executed: authentication, start of ciphering and optional equipment identity check.
(3) The SGSN requests the IP address of the GGSN from the domain name server with the APN. The SGSN then creates a tunnel identifier (TID) for the requested PDP context by combining the IMSI with the NSAPI. If the MS requests a dynamic address, then the SGSN sends a request to the GGSN for the allocation of a dynamic (temporary) IP address. The SGSN might not be able to fulfil the requested QoS, and may restrict the QoS attributes based on its capabilities and the current load. The SGSN sends a Create PDP context request message to the GGSN which includes:
  – the PDP type;
  – the PDP address (this field will be empty if a dynamic address is required);
  – the access point name (the GGSN may use the APN to find the required external network);
  – the negotiated QoS;
  – the TID.

The GGSN creates a new entry in its PDP context table and generates a charging Id (CID). The new entry allows the GGSN to route the data packets between the SGSN and the external PDP network, and to start producing charging data records (CDR). The GGSN may further restrict the QoS negotiated given its capabilities and the current load. The GGSN then returns a create PDP context response message to the SGSN containing:
- the TID;
- the PDP address (included if the GGSN has allocated a dynamic IP address);
- the negotiated QoS;
- the charging Id (CID).

(4) The SGSN inserts the NSAPI along with the GGSN IP address in its PDP context. If the MS has requested a dynamic address, the PDP address received from the GGSN is inserted in the PDP context. The SGSN returns an activate PDP context accept message to the MS containing:
- the PDP type;
- the PDP address (if a dynamic address is requested);
- the TI;
- the QoS negotiated.

The SGSN is now able to route the data packets between the GGSN and the MS, and to start producing charging data records (CDR) with the same CID as the CDR stored by the GGSN for this subscriber.

4.2.2   PDP Context Deactivation

The deactivation of a PDP context may be initiated by the MS, the SGSN or the GGSN. All these procedures are very similar to each other. In the following sequence, illustrated in Figure 4.11, the PDP context deactivation as initiated by the MS is described:

(1) The MS sends a deactivate PDP context request message with the transaction identifier (TI) to the SGSN.
(2) Security functions may be executed: authentication, start of ciphering and optionally the equipment identity check.
(3) The SGSN sends a delete PDP context request message containing the TID to the GGSN. The GGSN removes the PDP context and returns a delete PDP context

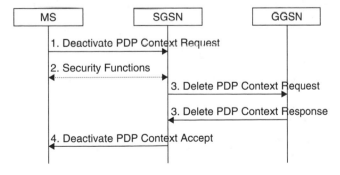

Figure 4.11   PDP context deactivation initiated by the MS

response message with the TID to the SGSN. If the MS was using a dynamic PDP address, then the GGSN releases this PDP address and makes it available for subsequent activation by another MS.

(4) The SGSN returns a deactivate PDP context accept message with the TI to the MS.

4.3 Packet Transfer Procedures

To start a GPRS application, the MS must be GPRS attached (in STANDBY or in READY state) and have activated a suitable PDP context.

The TLLI (temporary logical link identity) on the LLC layer and the NSAPI (network service access point identifier) on the SNDCP layer identify the PDP context between the MS and the SGSN. The TID (tunnel identifier) identifies the PDP context between the SGSN and the GGSN.

4.3.1 Mobile Originated Packet Transfer

When the MS wants to transmit data over the GPRS network, it must first send a packet channel request on a packet random access channel (PRACH) if this logical GPRS channel is configured in the cell, or on a random access channel (RACH) if PRACH is not implemented. The PCU sends back a packet immediate assignment message containing one or several uplink state flags USFs (if the MS may bundle several time slots, the MS may have a different USF for each individual time slot) and one temporary flow identifier TFI (the TFI is the same for all time slot), this message is sent over the radio interface on a packet access grant channel (PAGCH) if this logical GPRS channel is configured in the cell or with a access grant channel (AGCH) if PAGCH is not implemented.

Now, one or several UL time slots are assigned to the MS. As soon as the MS receives one of his USF on the corresponding DL time slot, the MS has to send some data with the TFI on the following UL radio block of this time slot.

4.3.2 Mobile Terminated Packet Transfer

If the MS is in STANDBY state, only the current routing area is known to the SGSN, so the MS must be paged by the SGSN before a DL data transfer is possible. The GPRS paging procedure is illustrated in Figure 4.12. and described in the following sequence.

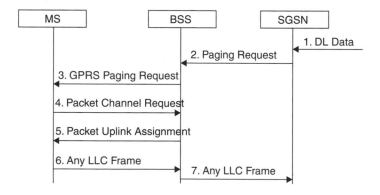

Figure 4.12   GPRS paging

(1) The SGSN receives some downlink data packets for an MS which is currently in STANDBY state.

(2) The SGSN sends a BSSGP paging request message (containing the P-TMSI, the routing area identity and the negotiated QoS for the PDP context) which initiates a paging procedure to the BSS serving the MS.

(3) One or several PCUs (it depends on whether the cells in the routing area are controlled by one or several PCUs) page the MS with a paging request message (containing the P-TMSI) in each cell belonging to the routing area concerned. Each message is sent over the radio interface on a packet paging channel (PPCH) if this logical GPRS channel is configured in the cell or on a paging channel (PCH) if PPCH is not implemented.

(4) Upon reception of a GPRS paging request message, the MS sends a packet channel request on a packet random access channel (PRACH) if this logical GPRS channel is configured in the cell or with a random access channel (RACH) otherwise.

(5) The PCU sends back a packet immediate assignment (one or several DL time slots are assigned to the MS) containing the temporary flow identifier (TFI), this message is sent over the radio interface with a packet access grant channel (PAGCH) if this logical GPRS channel is configured in the cell or with a access grant channel (AGCH) otherwise.

(6) The MS shall respond with any single valid LLC frame that will be implicitly interpreted as a paging response message by the SGSN. When responding, the MS changes MM state to READY.

(7) Upon reception of the LLC frame, the PCU adds the cell global identity (CGI) including the routing area code (RAC) and the location area code (LAC) of the cell and sends the LLC frame to the SGSN. The SGSN then considers the LLC frame to be an implicit paging response message.

Following the above sequence the MS is in READY state and, as soon as the MS 'recognizes' its own TFI on one of the DL time slots allocated to it, the MS 'knows' that it is the legitimate destination for this DL radio block, i.e. a packet transfer has started.

5

Changes in the Radio Subsystem for GPRS

5.1 Overview and Key Architecture

If GPRS is being introduced as part of the roll-out of a new GSM network, then presumably the necessary planning will have been included from the start. A scalable solution will have been developed to ensure that capacity increases (rather than modifications) can be carried out as necessary.

If, however, GPRS is being added to an existing network (the usual case), the extent of the modifications required to the BSS when implementing GPRS will depend on several factors:

- equipment obsolescence: can it be up-graded or must it be replaced?
- the anticipated traffic volume: is the existing Um capacity enough?
- whether the radio network is 'capacity' or 'coverage' limited: urban or rural areas?
- which coding schemes are being implemented: does Abis need upgrading?
- whether TS allocation is fixed or flexible: one or all carriers available for GPRS?
- whether radio resource (RR) allocation is prioritized: circuit switched before packet oriented?

5.1.1 Overview: Modifications to BSS

The best-case scenario:

- extra capacity is not needed (initially),
- only software upgrade required in BTS,
- PCU is added to BSC,
- only one SGSN is implemented,
- Gb interface realized via nailed up connections (nuc) from BSC via MSC to SGSN,
- Gn interface is realized via commercial IP based network,
- 7 TS reserved for GPRS on the same carrier as the BCCH (Broadcast Control Channel, see Section 5.1.3),
- dedicated resources, i.e. no flexible allocation of GSM/GPRS time slots.

GPRS Networks.   G. Sanders, L. Thorens, M. Reisky, O. Rulik, S. Deylitz
© 2003 John Wiley & Sons, Ltd.   ISBN: 0-470-85317-4

The worst-case scenario will vary according to the particular implementation:

- The long established networks will have significant numbers of older base stations in operation that cannot be upgraded to GPRS. In such cases, the purchase and commissioning of new base stations will be a major investment in terms of time, money and resources.
- Operators planning to introduce services such as i-mode, which use GPRS as a bearer, will be faced with the need for a significant capacity increase in urban areas to cope with the anticipated high demand.
- Operators wishing to introduce flexible allocation of time-slots for GSM or GPRS, will have to consider throughput models involving both circuit switched and packet switched traffic. In such cases the peak kbit s^{-1} for packet data will have to be converted to Erlang in order to enable capacity requirements and blocking probability to be calculated.
- All cases that result in significant Um capacity increases will also require more (or higher capacity) BSCs, and increased E1/T1 capacity on the Abis. If the Gb interface between the PCU and SGSN is realized via nailed up connections (nucs), then more capacity will also be required on the A interface between BSC and MSC.

5.1.2 Functional Split Between CCU and PCU for GPRS

As mentioned previously, the implementation of GPRS requires new BSS network entities:

- Channel codec units (CCU),
- Packet control units (PCU).

As illustrated in Figure 5.1, the CCU is realized as software installed in each BTS, while the PCU is hardware that can be implemented at the BTS, BSC or SGSN site.

The CCU performs the following functions:

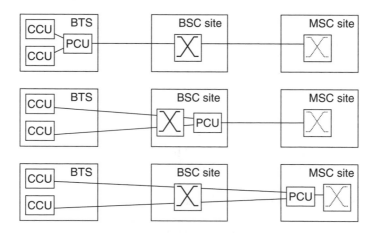

Figure 5.1 Placement of CCU and PCU

- channel coding functions using coding schemes 1 to 4, including forward error correction and interleaving;
- radio channel measurement functions, including received quality level (RxQual), received signal level (RxLev), and values related to timing advance (TA).

The PCU is responsible for the following functions:

- radio channel management functions, including power control (PC), congestion control, and broadcast control information;
- LLC layer PDU segmentation into RLC blocks (downlink transmissions);
- LLC layer PDU reassembly from RLC blocks (uplink transmissions);
- PDCH (packet data channel, see Section 5.1.3) scheduling functions for UL and DL data transmission;
- ARQ (automatic repeat request) functions:
- PDCH uplink, including acknowledged and unacknowledged RLC blocks;
- PDCH downlink, including buffering and retransmission of RLC blocks;
- channel access control functions, including access requests and grants.

At the time of writing, commercial GPRS networks known to the author have the PCU positioned at the BSC. It is clear from the above that the functional split between the CCU and PCU is not the same as that between the BTS and BSC. What is not so clear is that, because the GPRS specifications allow a certain functional flexibility with regard to *how* the CCU and PCU are realized, the CCU can control some of the PCU functionality and vice versa. This text explains the functional split as described in the specifications and implementations known to the author. However it must be said that variation can and probably will exist.

In GSM, power control (PC) for each mobile is controlled in the BTS, but in GPRS the CCU (in the BTS) takes measurements for PC and sends the values to the PCU. In GSM, each time slot is used by one mobile, so the timing advance (TA) for the mobile's UL transmissions can be continuously monitored and adjusted, but in GPRS, up to eight mobiles can share an UL time slot and they can use it at irregular intervals which means the TA for each of these mobiles must be controlled individually.

In Chapter 1 the concept of a 'multi-frame' was introduced. For traffic channels in GSM, we have a multi-frame structure of 26 TDMA frames. Of these 26 TDMA frames, the first twelve are used for traffic, the thirteenth for signalling (PC and TA), the following twelve for traffic again, and the last one is an idle frame (when half rate is used, the last frame contains signalling for the second mobile). In GPRS a new multi-frame structure has been introduced for the packet traffic channel, which consists of 52 TDMA frames. The new multi-frame is illustrated in Figure 5.2.

To understand the multi-frame structure, the concept of a 'radio block' must be understood. User information is coded in blocks of 456 bits known as 'radio blocks'. A time slot can only carry 114 bits of user information, so it takes four time slots to transmit a radio block. So when a subscriber is allocated the use of a time slot, the MS will transmit the radio block in the allocated time slot in four consecutive TDMA frames. After the radio block has been sent, the same user may again be allocated use of the time slot or another subscriber may receive it. This means that when subscribers share a time slot,

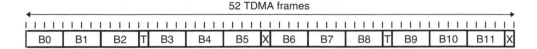

52 TDMA frames

X = idle frame
T = frame used for PTCCH
PTCCH = Packet timing advance control channels
B0–B11 = radio blocks

Figure 5.2   The GPRS multi-frame structure

the current user can only be changed after every four time slots. It also means that the coding scheme can also only be changed every four time slots, i.e. a coding scheme is applied to an entire radio block not only to a time slot.

The first twelve TDMA frames of a multi-frame are for traffic and they are divided into three radio blocks (three blocks of four time-slots). After these three radio blocks, one TDMA frame is reserved for signalling. This sequence of twelve TDMA frames for traffic (three radio blocks) and one TDMA frame for signalling, is repeated four times to produce the complete GPRS multi-frame. This means that in each GPRS multi-frame, there are four TDMA frames for signalling and two of these four are reserved for the continuous timing advance update procedure which is carried on the packet timing advance control channel (PTCCH). This PTCCH is shared by up to 16 subscribers, so we must consider a structure of eight GPRS multi-frames, as represented in Figure 5.3, to allow for the UL and DL transmission of the 16 possible subscribers.

As part of the packet resource assignment, each individual MS is assigned a timing advance index (TAI), which specifies the PTCCH sub-channel that the MS should make use of. The MS sends access bursts on the sub-channel defined by the TAI (0...15) on the PTCCH. These access bursts received on the uplink PTCCH are used by the CCU (in the BTS) to derive the timing advance. The new timing advance values are sent via a downlink signalling message (TA message) on the downlink PTCCH. The BTS updates the timing advance values in the next TA message following the access burst. This is illustrated in Figure 5.3, which shows the mapping of the uplink access bursts and downlink timing advance signalling messages. An interesting point is that the four mobiles that have TAI = 0, 1, 2, or 3 (transmitting an access burst in frames numbered 0, 2, 4, 6) all receive their updated timing advance value in TA message 2, i.e. the four TA values are interleaved across frames 8, 10, 12, and 14. This is similarly true for TAI = 4, 5, 6, 7 and TA message 3, and so on.

When the MS receives the updated value of TA from the CCU (in the BTS) on the downlink PTCCH, it shall always use the last received TA value for the uplink transmission of normal bursts.

5.1.3 GPRS Signalling

One aspect of the air interface which often causes confusion is the difference between the *physical* and the *logical* organization of the air interface.

The physical organization of the radio interface, as previously described, is relatively easy to understand. One physical channel is determined by a frequency band of 200

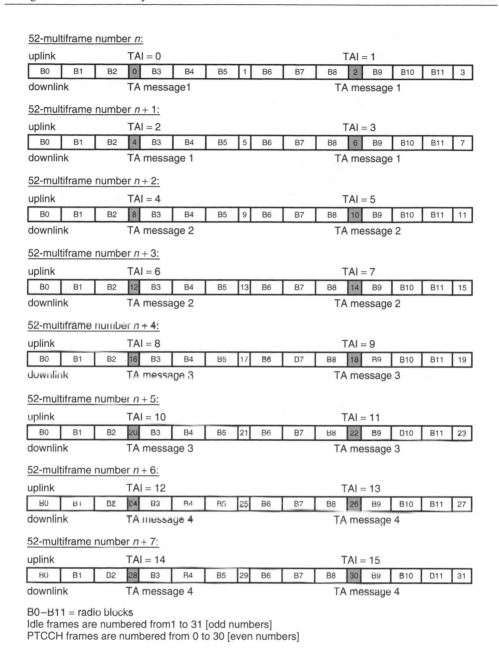

Figure 5.3   Mapping of the uplink access bursts and downlink timing advance signalling messages

kHz and a time slot with a duration of 577 µs. This physical channel may be uplink (between 880 and 915 MHz for GSM900, between 1710 and 1785 MHZ for GSM1800) or downlink (between 925 and 960 MHz for GSM900, between 1805 and 1880 MHZ for GSM1800). These frequency bands are distributed to the base stations in specific groups known as 'clusters' such that the interference between cells using the same frequencies is minimized. Each frequency is divided in the time domain into time slots. The eight time slots on any given frequency are known as a 'TDMA frame'.

The logical organization of the radio interface is somewhat more complex. In GSM and GPRS, one physical channel can have several logical channels multiplexed onto it, e.g. the physical channel allocated to a GPRS user can be used for transporting data traffic, signalling, and, in some configurations, even for paging. This logical organization is described in the following two sections – GSM in the first one and GPRS in the second one. These sections will describe such things as: which time slot and frequency band the application (speech or data application) of a user is transported on, which time slot and frequency band can a subscriber send a channel request on, and which time slot and frequency band may a subscriber be paged on. It should be noted that several new logical channels have been specified for GPRS but where these are not available, the GPRS signalling will be carried out over the normal GSM channels.

5.1.3.1 GSM Logical Channels

The logical channels defined for GSM are illustrated in Figure 5.4. They can be separated into two categories: the traffic channels and the control channels.

Traffic Channels

The traffic channels (TCHs) are intended to carry either speech or data traffic in circuit switched mode. The full rate (FR) TCHs have a gross bit rate of 22.8 kbit s^{-1} and the half rate (HR) TCHs have a gross bit rate of 11.4 kbit s^{-1} (a subscriber having a HR TCH only uses every second time slot, i.e. a frequency carrier with HR subscribers has double the capacity as that with FR subscribers). The net bit rate depends on the reliability of the channel coding: if the net bit rate is high, then the redundancy is small; if the net bit rate is small, then the redundancy is high.

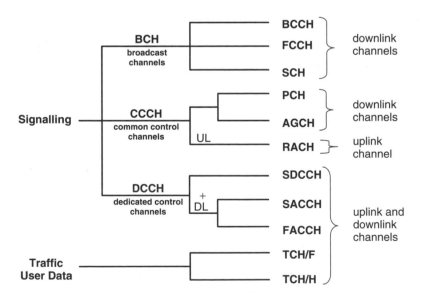

Figure 5.4   GSM logical channels

The following traffic channels are defined for speech:

- full rate traffic channel for speech (TCH/FS), whose net bit rate is 13 kbit s^{-1};
- half rate traffic channel for speech (TCH/HS), whose net bit rate is 5.6 kbit s^{-1};
- enhanced full rate traffic channel for speech (TCH/EFS), whose net bit rate is 12.2 kbit s^{-1}.

TCH/FS was introduced at the beginning of GSM as part of Phase 1 and its speech quality is satisfactory. TCH/HS was introduced at Phase 2 of GSM, and is designed to increase the capacity of the air interface by allowing two users to share a voice channel. HR was also the first step towards adaptive multirate (AMR switches actively between EFR, FR and HR depending on the quality of the air interface). With a good quality air interface there is little perceptible difference between FR and HR. HR speech quality deteriorates noticeably with decreasing quality of the air interface. This is due to the low amount of speech data actually transmitted (even though the transmission itself is highly redundant). The Phase 2+ of GSM defined TCH/EFS, which delivers speech quality that may be even better than that of TCH/FS, especially for high-pitched voices, when the quality of the air interface is good.

The following traffic channels are defined for data:

- full rate traffic channel for 14.4 kbit s^{-1} user data (TCH/F14.4);
- full rate traffic channel for 9.6 kbit s^{-1} user data (TCH/F9.6);
- full rate traffic channel for 4.8 kbit s^{-1} user data (TCH/F4.8);
- full rate traffic channel for $\leqslant$ 2.4 kbit s^{-1} user data (TCH/F2.4);
- half rate traffic channel for 4.8 kbit s^{-1} user data (TCH/H4.8);
- half rate traffic channel for $\leqslant$ 2.4 kbit s^{-1} user data (TCH/H2.4).

The 14.4 kbit s^{-1} channel coding listed above was only introduced in GSM Phase 2+ for HSCSD services. All the others were defined at the introduction of GSM. In Phases 1 and 2, a subscriber could not bundle several time slots for his data application (e.g. a WAP session), therefore the maximal bit rate was 9.6 kbit s^{-1}.

As stated previously, these traffic channels are structured inside multi-frames of 26 TDMA frames. Of these 26 TDMA frames, the thirteenth is used for signalling. This is an example of different logical channels being sent on the same physical channel. First the time slot carries a traffic channel (TCH) then, in the thirteenth TDMA frame, it carries a so called 'slow associated control channel' (SACCH – see next section), which is used for signalling information. The twenty-sixth is either idle if it is a full-rate channel, or is used for signalling (SACCH) for the second subscriber if it is a half-rate channel.

Not all time slots are organized using traffic multi-frames of 26 TDMA frames. Some time slots must be reserved and only used for signalling (see next section). The allocation of such time slots follows a multi-frame comprising 51 TDMA frames. Various combinations of logical channels have been standardized for use within this signalling multi-frame. The combination used will depend on the configuration of the BTS and the requirements of the network operator.

Control Channels
The control channels are intended to carry signalling or synchronization data. Three categories of control channels are defined: broadcast, common, and dedicated control channels.

The broadcast channels are DL only and carry information for all subscribers inside the cell. Three specific broadcast channels are defined:

- Frequency correction channel (FCCH): this channel sends a signal with a constant frequency (i.e. a pure carrier wave) such that the MS is able to synchronize itself with the BTS in the frequency domain.
- Synchronization channel (SCH): this channel sends a signal with a long fixed-bit sequence (training sequence) such that the MS is able to synchronize itself with the BTS in the time domain, i.e. to allow for the DL transmission delay. This channel also sends the base transceiver station identity code (BSIC) and the values of two counters which tell the mobile what the current TDMA frame numbers are for the two multi-frames, i.e. with these two values the MS knows which logical channel is being sent/received on the traffic multi-frame and on the signalling multi-frame.
- Broadcast control channel (BCCH): this channel contains parameters used by the MS to access the network, e.g. cell global identity (CGI), location area identity (LAI), routing area identity (RAI), frequency hopping algorithm, channel combination, cipher mode (if the information is ciphered or not over the radio interface), support of HSCSD, GPRS, EDGE (if HSCSD, GPRS, EDGE are supported or not).

The term 'channel combination' (above) refers to which combination of logical channels are implemented in the cell and is described in detail below.

If GPRS is supported, then two possibilities exist for the transport of GPRS signalling. Either no new logical channels are defined and the GPRS signalling is transported by the GSM logical channels (see Figure 5.4) or new GPRS logical channels are implemented (see Figure 5.5). In the latter case, a PBCCH (packet broadcast control channel) may exist, therefore the BCCH broadcasts details of the physical channel that is carrying this PBCCH.

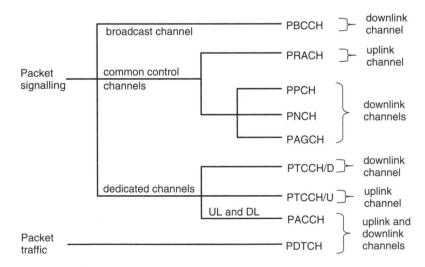

Figure 5.5 GPRS logical channels

The common control channels are unidirectional (either UL only or DL only) and are used for the initial access:

- Paging channel (PCH): DL only, to search for (to page) the MS in all cells of the current location area in the case of an incoming circuit switched call, or in all cells of the current routing area in the case of an incoming packet switched service.
- Random access channel (RACH): UL only, the MS requests the allocation of a stand-alone dedicated control channel (SDCCH, see below).
- Access grant channel (AGCH): DL only, the network allocates a SDCCH to the MS.
- Notification channel (NCH): DL only, the network notifies all mobile stations of a specific group of a voice broadcast service (VBS) or a voice group call service (VGCS).

The dedicated control channels are bidirectional (UL and DL) and are used for dedicated signalling:

- Stand-alone dedicated control channel (SDCCH): for call set-up, authentication, cipher start, temporary mobile station identity (TMSI) reallocation (allocation of a new temporary identifier for the MS), international mobile subscriber identity (IMSI) attach/detach (the MS sends a message that it is switched on/off), location update procedure, and also the transport of short message service (SMS).
- Slow associated control channel (SACCH): allocated together with a SDCCH or a TCH, a SACCH contains control information for the maintenance of a connection, e.g. power control and timing advance on the DL, and measurement reports on the UL.
- Fast associated control channel (FACCH): allocated instead of a TCH in case signalling messages need to be exchanged between BTS and MS, e.g. during handover and call release, signalling data instead of voice data are transmitted on the time slot allocated to the MS.

If the BTS has only one or two carriers, only one time slot must be reserved for signalling. This signalling channel is often referred to as 'time slot zero' as it is found on the first time slot (TS0) of the carrier C0. In Figure 5.6, we can see that all of the necessary logical channels are present in one multi-frame of 51 TDMA frames:

- In DL, all broadcast control channels (FCCH, SCH and BCCH), all DL common control channels (PCH, AGCH and maybe NCH) and all DL dedicated control channels (SDCCH and SACCH). These dedicated control channels are allocated in blocks of four consecutive time slots, i.e. in radio blocks. Up to four subscribers may individually receive a dedicated channel on this multi-frame.
- In UL, all possible RACH (the only UL common control channel) and all UL dedicated control channels (SDCCH and SACCH).

If the cell has more carriers (three or more), then there will be more subscribers and more traffic, therefore more signalling channels will be necessary. The addition of more signalling channels is done in two or three steps:

DL: FCCH + SCH + BCCH + PCH + AGCH + SDCCH/4 + SACCH/4

| F | S | BCCH | CCCH | F | S | CCCH | CCCH | F | S | SDCCH 0 | SDCCH 1 | F | S | SDCCH 2 | SDCCH 3 | F | S | SACCH 0 | SACCH 1 | I |
| F | S | BCCH | CCCH | F | S | CCCH | CCCH | F | S | SDCCH 0 | SDCCH 1 | F | S | SDCCH 2 | SDCCH 3 | F | S | SACCH 2 | SACCH 3 | I |

UL: RACH + SDCCH/4 + SACCH/4

| SDCCH 3 | R | R | SACCH 2 | SACCH 2 | RRRRRRRRRRRRRRRRRRRRRRRR | SDCCH 0 | SDCCH 1 | R | R | SDCCH 2 |
| SDCCH 3 | R | R | SACCH 0 | SACCH 1 | RRRRRRRRRRRRRRRRRRRRRRRR | SDCCH 0 | SDCCH 1 | R | R | SDCCH 2 |

F: FCCH   S: SCH   CCCH: PCH or AGCH   I: Idle   R: RACH

Figure 5.6   Signalling in a cell with one or two carriers

DL: FCCH + SCH + BCCH + PCH + AGCH

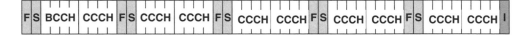

UL: RACH

F: FCCH   S: SCH   CCCH: PCH or AGCH   I: Idle   R: RACH

Figure 5.7   Basic signalling in a cell with several carriers

(1) On the time slot 0 of the carrier C0, we implement a multi-frame with more common control channels (for initial access) but without the dedicated control channels (Figure 5.7).
(2) On the time slot 0 of another carrier, the carrier C1, we have a multi-frame with all the dedicated control channels. Up to eight subscribers may individually receive a dedicated channel on this multi-frame (Figure 5.8).
(3) If there are not enough common control channels on the time slot 0 of the carrier C0, then such new channels are defined on the time slot 2 of this carrier C0. If even more channels are required (i.e. the BTS has many carrier frequencies), then other new channels may be defined on time slot 4 and also on time slot 6. These new common control channels are defined in similar multi-frames of 51 TDMA frames.

DL: SDCCH/8 + SACCH/8

SDCCH 0	SDCCH 1	SDCCH 2	SDCCH 3	SDCCH 4	SDCCH 5	SDCCH 6	SDCCH 7	SACCH 4	SACCH 5	SACCH 6	SACCH 7	▓
SDCCH 0	SDCCH 1	SDCCH 2	SDCCH 3	SDCCH 4	SDCCH 5	SDCCH 6	SDCCH 7	SACCH 0	SACCH 1	SACCH 2	SACCH 3	▓

UL: SDCCH/8 + SACCH/8

SACCH 5	SACCH 6	SACCH 7	▓	SDCCH 0	SDCCH 1	SDCCH 2	SDCCH 3	SDCCH 4	SDCCH 5	SDCCH 6	SDCCH 7	SACCH 4
SACCH 0	SACCH 1	SACCH 2	▓	SDCCH 0	SDCCH 1	SDCCH 2	SDCCH 3	SDCCH 4	SDCCH 5	SDCCH 6	SDCCH 7	SACCH 0

Figure 5.8 Additional signalling in a cell with several carriers

(I)   TCH/F + FACCH/F + SACCH/F

(II)   TCH/H(0,1) + FACCH/H(0,1) + SACCH/H(0,1)

(III)   TCH/H(0) + FACCH/H(0) + SACCH/H(0) + TCH/H(1) + FACCH/H(1) + SACCH/H(1)

(IV)   FCCH + SCH + CCCH + BCCH

(V)   FCCH + SCH + CCCH + BCCH + SDCCH/4 + SACCH/4

(VI)   CCCH + BCCH

(VII)   SDCCH/8 + SACCH/8

Figure 5.9 Logical channel combinations I to VII

If Figures 5.6, 5.7 and 5.8 are compared with each other, it can be seen that the 51 multi-frame can contain various combinations of the logical channels defined for GSM. There are actually seven different combinations of logical channels. As described above and illustrated in Figure 5.9, a cell with one or two carriers will use 'Combination-V' while a cell with three or four carriers might make use of 'Combination-IV' on TS0 and 'Combination-VII' on another time slot.

5.1.3.2 GPRS Logical Channels

A certain number of GPRS logical channels may be implemented which have similar tasks to those of GSM. Packet data traffic channels (PDTCH) are intended to carry user data in packet switched mode. As such a channel may be shared by several subscribers, the mobile 'owner' of the DL radio block shall be indicated by the temporary flow identifier (TFI), and also in each DL radio block, the uplink state flag (USF) is sent, indicating which subscriber can send next in the UL radio channel. The packet broadcast control channel (PBCCH) broadcasts parameters used by the MS to access the network for packet transmission operation. In addition to those parameters, the PBCCH reproduces the information transmitted on the BCCH to allow circuit switched operation, such that a MS in GPRS attached mode monitors only the PBCCH (if it exists). The existence and the location of the PBCCH in the cell are indicated on the BCCH, i.e. the MS will first make the frequency and time adjustments using the FCCH and the SCH, then consider the information on the BCCH, and with this information will find the location of the PBCCH. In the absence of PBCCH, the BCCH is used to broadcast information for packet operation.

Of the many parameters contained in the PBCCH, the following four are important for the organization of the GPRS logical channels:

- BS_PBCCH_BLKS (1, ..., 4) indicates the number of blocks allocated to the PBCCH in the multi-frame. This parameter is broadcast on PBCCH in the first block, the block B0 (see Figure 5.2, GPRS multi-frame structure).
- BS_PCC_CHANS indicates the number of physical channels (i.e. time slots) carrying packet common control channels (PCCCH), including the physical channel carrying the PBCCH. The PBCCH, in addition, indicates the physical location of those channels.
- BS_PAG_BLKS_RES indicates the number of blocks on each physical channel carrying PCCCH per multi-frame, where neither PPCH nor PBCCH should appear (i.e. where PRACH in UL and PAGCH in DL may appear).
- BS_PRACH_BLKS (0, ..., 12) indicates the number of blocks reserved in a fixed way to the PRACH channel on any physical channel carrying PCCCH.

Similarly to GSM, the following packet common control channels are defined: the packet paging channel (PPCH), the packet random access channel (PRACH), the packet access grant channel (PAGCH) and the packet notification channel (PNCH). If no such channels are implemented, the information for packet switched operation is transmitted on the GSM common control channels.

Finally, the following packet dedicated control channels are defined for GPRS:

- The packet associated control channels (PACCH): the PACCH conveys signalling information related to a given MS. The signalling information includes, for example, acknowledgements and power control information. PACCH also carries resource assignment and reassignment messages. The PACCH shares resources with the PDTCHs that are currently assigned to one MS. Additionally, an MS that is currently involved in packet transfer can be paged for circuit switched services on PACCH.
- The packet timing advance control channel (PTCCH): for the continuous timing advance update procedure described in Section 5.1.2.

All these packet data logical channels are organized in multi-frames of 52 TDMA frames (see Figure 5.2).

5.2   Introduction of EDGE, ECSD and E-GPRS

5.2.1   General Description

EDGE is a new technology that offers much higher data rates over the air interface. EDGE stands for enhanced data rates for GSM evolution, and can be combined with both

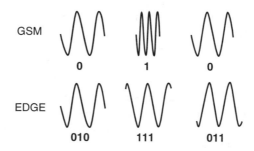

Figure 5.10 Comparison of GSM and EDGE modulation

HSCSD and GPRS to realize new coding schemes which, along with time slot bundling, can theoretically provide data rates around three-times higher than 'normal' GPRS. EDGE can be integrated into existing RSS and can be thought of as an upgrade for the modulation of bits onto the air interface.

The term 'modulation' refers to how the bits are transformed into radio waves and sent across the air interface. The so called 'GMSK' modulation used in GSM sends the bits one at a time, with EDGE however the so called '8PSK' modulation sends the bits three at a time. Hence EDGE offers roughly triple the bit rate of GSM.

A comparison of GSM and EDGE modulation is given in Figure 5.10. With GSM, a single bit (0 or 1) is represented by a higher or a lower frequency. With EDGE a group of three bits (001, 010, 011, ..., 111) is represented by one of eight different phase values, transmitted with a constant frequency.

The gross data rate over the radio interface with EDGE is 69.2 kbit s^{-1}, and several new coding schemes are introduced which offer net bit rates of up to 59.2 kbit s^{-1}. If a subscriber has all eight time slots of a carrier, the maximal theoretical data rate with EDGE is then:

$$59.2 \text{ kbit s}^{-1} \times 8 \text{ time slots} = 473.6 \text{ kbit s}^{-1}.$$

As EDGE offers such high data rates, this standard is also considered as belonging to the third generation. At the time of writing, several operators in the USA are implementing EDGE instead of UMTS or CDMA2000.

In the network, new modules must be implemented in the base stations, such that the new modulation is supported. Some other software updates must be done (in the PCU and in the SGSN for example) to support the additional signalling requirements. Another important remark regarding the implementation of the network, and especially of the BSS, is that more bandwidth (around three-times more) is necessary in the network, and especially on the Abis interface, because of the higher transmission rates over the radio interface. It is very common in GSM to install some chains of BTSs that are connected to the BSC, but because of the higher information load the length of these chains may need to be shortened, otherwise the link between the BSC and the first BTS has insufficient bandwidth to transport the volume information.

Of course, new mobiles supporting the new modulation are also necessary.

Apart from the new modulation, EDGE uses the same radio interface as GSM. Furthermore, EDGE can transport either circuit or packet switched applications. For these, two different applications are distinguished: ECSD: enhanced circuit switched data; and E-GPRS: enhanced general packet radio service.

5.2.2 ECSD

In Figure 5.11 the different coding schemes for ECSD are presented. At the beginning of GSM, the first three were introduced (for full rate). At that time, the subscriber could have only one time slot in each TDMA frame. The modulation is GMSK (Gaussian minimum shift keying) which interprets a bit from one of two possible frequency values.

Schemes	Modulation	Net data rate (kbit s^{-1})
TCH/F2.4	GMSK (gross data rate = 22.8 kbit s^{-1})	2.4
TCH/F4.8		4.8
TCH/F9.6		9.6
TCH/F14.4		14.4
E-TCH/F28.8	8PSK (gross data rate = 69.2 kbit s^{-1})	28.8
E-TCH/F32.0		32.0
E-TCH/F43.2		43.2

Figure 5.11   The seven coding schemes for ECSD

Schemes	Modulation	Net data rate (kbit s^{-1})
MCS-1	GMSK (gross data rate = 22.8 kbit s^{-1})	8.8
MCS-2		11.2
MCS-3		13.6 or 14.8
MCS-4		17.6
MCS-5	8PSK (gross data rate = 69.2 kbit s^{-1})	22.4
MCS-6		27.2 or 29.6
MCS-7		44.8
MCS-8		54.4
MCS-9		59.2

Figure 5.12   The nine coding schemes for E-GPRS

After that, high speed circuit switched data (HSCSD) was introduced with the coding scheme TCH/F14.4. With this service, the subscriber may bundle several time slots, actually up to four, because the MSC can not handle an application with a transmission rate higher than 64 kbit s^{-1}. HSCSD is suitable for real-time applications.

With ECSD, the subscriber may use the last three coding schemes. The modulation is 8PSK (phase shift keying with eight values), the new modulation with eight different phase values for coding groups of three bits. One real advantage is that the subscriber does not need to bundle a lot of time slots to get a high transmission rate, and so needs fewer resources over the radio interface. Some examples are 43.2 kbit s^{-1} with one time slot, $2 \times 28.8 = 57.6$ kbit s^{-1} with two time slots, or even $2 \times 32 = 64$ kbit s^{-1} with two time slots.

5.2.3 E-GPRS

In Figure 5.12, the different coding schemes for E-GPRS are presented. There are four coding schemes with the 'old' modulation and five with the 'new' modulation. With the E-GPRS services, the subscriber has a connection with the packet oriented core network, i.e. the SGSN and the GGSN. Therefore, there is no limitation on the transmission rate (other than that imposed by QoS parameters in the network). The subscriber may bundle up to eight time slots of one carrier, so the maximal theoretical transmission rate is $8 \times 59.2 = 473.6$ kbit s^{-1}.

6

Core Network

One of the main concerns during the development of the GPRS standard was that the impact on existing GSM networks should be kept to a minimum. As a result, the additional core network elements (mainly the serving and the gateway GPRS support nodes, SGSN and GGSN) and associated interfaces, can be thought of as an 'overlay' to the existing GSM infrastructure. In addition to simplifying the roll-out, this also means there should be little or no down time (lower quality of service for GSM users) during implementation.

In a GPRS equipped GSM network, there is a switching network for the circuit switched traffic known as the Network switching subsystem (NSS), and a routing network for the packet oriented traffic (GPRS support node network for data, GSN network), which together make up the core network.

The core network is strictly separated from the radio side (base station system) and it comprises:

- the interconnected network of exchanges, supporting databases and interfaces, for circuit switched traffic (such as voice, circuit switched data, SMS, signalling), and
- the interconnected network of support nodes, routers, supporting databases and interfaces, for the transfer of packet oriented traffic (IP-based traffic such as web access, e-mail, file transfer).

This separation allows for a smooth evolution from GSM through GPRS to UMTS, as each part of the core network (CS and PO) can interface with different radio access technologies, i.e. GSM via GERAN (BTS and BSC); UMTS via UTRAN (Node-B and RNC).

Furthermore, the GPRS support node (GSN) network is completely separated from the existing GSM core network, with the exception of certain interfaces that enable certain common functions such as authentication, transfer of subscriber data, and mobility management. To a large extent, the build up and the design of the GSN network is independent of the existing NSS, which is shown in Figure 6.1. Separate connections from the BSCs to the GSN network are installed in addition to the connections to the NSS. These units, which are added to the BSC to enable a connection to the GSN, are called packet control units (PCU).

GPRS Networks. G. Sanders, L. Thorens, M. Reisky, O. Rulik, S. Deylitz
© 2003 John Wiley & Sons, Ltd. ISBN: 0-470-85317-4

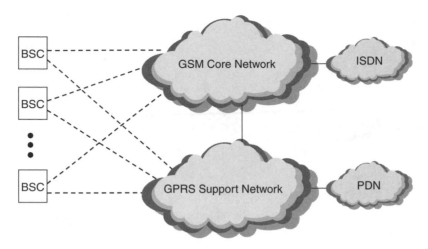

Figure 6.1 Physical separation of radio, GSM core and GPRS core networks

The implementation of new network elements means that new interfaces are necessary. This, along with new functionalities (such as session management and common mobility management), means that the impact on GSM elements, such as the HLR and VLR, takes the form of additions rather than fundamental changes.

The main elements of the packet switched core network are illustrated in Figure 6.2 and consist of the following:

- serving GPRS support node (SGSN), which is the unit that connects the radio network with the core network;
- gateway GPRS support node (GGSN), which is the unit that connects the external data network (intra-/Internet) with the core network;

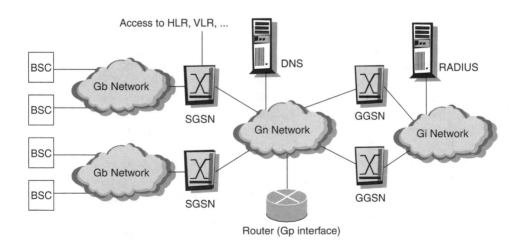

Figure 6.2 Components of the GPRS core network

- access network for interconnecting the PCUs and the SGSNs (in the following text referred to as the Gb network);
- core network for interconnecting the SGSNs and GGSNs (in the following text referred to as the Gn network);
- additional elements in the Gn network, such as the DNS server for address resolution or routers to provide interconnectivity with the GPRS support node networks of other providers (Gp interface);
- additional elements connected to the GGSN via the Gi interface, such as a RADIUS server for authentication, authorization and accounting or to administer an IP-address pool;
- connections from elements of the GSN (mainly the SGSN) to elements of the circuit switched core network like the HLR, the VLR or the SMS centre.

Figure 6.2 is only intended to provide an overview of the most common elements and interfaces. Detailed information will be provided in the following text and further details can also be found in the Chapter 3.

6.1   Serving GPRS Support Node (SGSN)

The SGSN occupies the same hierarchical level as an MSC in the circuit switched core network and, as can be seen from the following list, handles many functions comparable to a VMSC. In fact, the SGSN contains a register called the SGSN location register (SLR), which stores subscriber-related information similar to the information stored by the VLR in the MSC, i.e. the MSC/VLR pair is analogous to the SGSN/SLR pair.

The following functions are performed by the SGSN:

- serve all GPRS mobile stations within a certain area (an SGSN connects to a certain set of PCUs, each serving a certain set of routing areas – a routing area can be thought of as a set of cells);
- mobility management functions (attach, detach, routing area update (RAU) and paging) for user registration and updating the data stored in the SLR;
- store and maintain subscriber information in the SGSN location register (SLR) of all subscribers registered in the routing areas belonging to that particular SGSN;
- session management functions (PDP context activation/deactivation to establish a connection to a certain GGSN);
- packet handling functions (forwarding incoming user data from a PCU to a GGSN and vice versa, priority switching in the case of quality of service);
- control inter-SGSN routing area update (the new SGSN where the subscriber has registered contacts the old SGSN to receive the SLR entries);
- SMS handling (the SMS centre can be directly connected to the SGSN via the Gd interface);
- charging data collection (charging data records can be collected about the amount of data transferred in order to generate files for billing);
- prepaid handling (debiting of prepaid accounts for GPRS usage);
- performance management (to measure the traffic within this part of the network);
- maintenance and fault management (to discover problems during transmission and procedures);

- realizes protocols and interfaces (Gb to connect to PCU; Gn to connect to GGSN, other SGSN or DNS; Gr to connect to HLR; Gs to connect to MSC/VLR); these interfaces are implemented with modules running protocols such as BSSGP, LLC, SNDCP, GMM, SM and GTP.

With regard to the first point on the list, the area served by an SGSN is implementation dependent. It is possible to serve an entire national network with a single SGSN, in which case the area served by the node will be the entire country (all routing areas of the network). As the volume of traffic increases, it is possible to introduce more nodes that serve different geographical areas. This means that the geographical area served by one SGSN will be scaled down as the capacity of the network (and the number of nodes) is scaled up.

The physical realization of the SGSN varies between manufacturers. However, there is a certain commonality which arises primarily from the necessity to realize the above listed functions and transfer the protocol structures into hardware and software. For the purposes of this book, we have produced our own generic SGSN incorporating the elements most important for packet data transfer. We do not explicitly mention elementary units such as power supplies, clock generators, alarm collectors, etc.

Figure 6.3 illustrates what we can expect to find in a typical SGSN from any manufacturer. Some (or even all) functions shown as separate logical modules might be combined into one hardware unit, others might be separated. The interface cards are shown on the left side, the processor cards (realizing the protocols used in GPRS) on the right side. All modules are able to communicate via the internal ATM switching network.

The functions of the modules are:

- *MM*: performs mobility management functions (such as attach, detach, paging, routing area update) and contains the semipermanent subscriber data in the SLR. Also handles protocols for signalling exchange based on SS7 (e.g. with HLR).
- *PD/SH and GTP, packet dispatching and session handling*: handles all subscriber dependent packet/protocol processing such as session management (PDP context activation/deactivation), resource management, parts of the BSSGP, LLC, and SNDCP protocols necessary for the Gb and the Gn interface, and enables data packets to be tunnelled across the Gn interface (GTP).
- *ACC*: carries out statistical and accounting functions and stores billing data, which can be retrieved by an accounting and billing centre.
- *SLT*: handles the lower levels of SS7 signalling such as global title translation (GTT) or message transfer part (MTP) for the SS7 message routing for communication with entities such as HLR or VLR.
- *OAM*: realizes all operation, administration and maintenance functions to operate the system, e.g. remote configuration/update of SGSN software, alarm notification and fault clearance.
- *Interface Cards*: implement the protocols of layers 2 and 1, such as frame relay/E1 for the Gb interface, Ethernet 100BaseT or ATM/STM-1 for the Gn interface and PCM30 for Gr, Gs and Gd interfaces.

Seen from a hardware point of view, the protocol layers of a certain interface may be implemented in several hardware units. The Gb and Gn interfaces (illustrated in

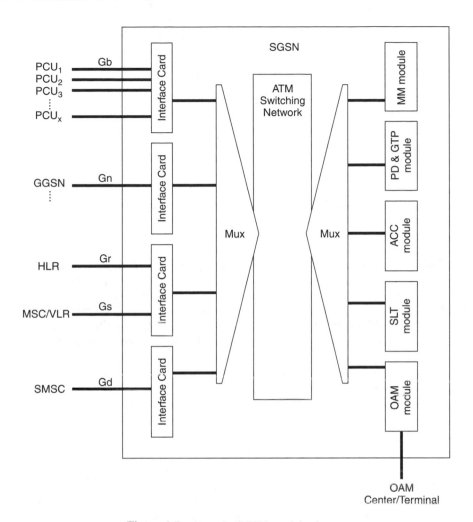

Figure 6.3 Generic SGSN modular layout

Figure 6.4) are good examples of this and are described in the following text. The first two layers E1 and frame relay are implemented in the hardware of the frame relay card. This card forwards the frames via the switching network to the packet dispatcher processor card that is running the higher layer protocols BSSGP, LLC and SNDCP. To transmit the user data to the corresponding GGSN, it must first be encapsulated in a GTP frame. The two separate cards for SNDCP and GTP would communicate via the internal ATM switching network. The tunnelling protocol (GTP) as well as UDP, TCP (not implemented yet) and IP can be implemented in the same hardware as the packet dispatching module or in a separate card. Finally, the UDP (or TCP) datagram is encapsulated in an IP packet, which needs to be forwarded to the interface card that provides the layer 2 and 1 functionality of the Gn interface, e.g. 100BaseT Ethernet or ATM via STM-1.

Since the SGSN should be protected against failure and down time which would result in data loss, it is necessary to establish redundancy mechanisms within the network

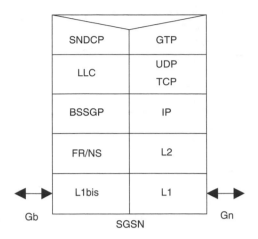

Figure 6.4   Protocol stack of the SGSN for data from the Gb to the Gn interface

elements. The need for redundancy usually means that each module will have a 'back-up' running in synchronous standby mode (the so called 'hot standby'); however some solutions have modules with their own internal redundancy. The common abbreviations '1 + 1' and 'n + 1' redundancy mean 'every module has a back-up' or 'several modules have one back-up' respectively. Load sharing is another common method of introducing redundancy and reducing the load on individual units. In which case two completely separate hardware units each process a percentage of the incoming traffic. This means that when a hardware failure occurs, the capacity of the SGSN is reduced by 50% until the fault can be cleared.

The choice of SGSN implementation strategy will also have an impact on the GSM core network. The different strategies for linking the SGSN(s) with the PCUs will be covered later but it is worth noting here that using 'nailed up connections' to implement the Gb interface results in a loss of capacity for GSM. The use of an external frame relay network to implement the Gb interface has no such consequences but needs the installation of additional elements and incurs additional expense.

6.2   Gateway GPRS Support Node (GGSN)

The GGSN is the gateway between the GPRS mobile and external packet data networks (PDN). The GGSN 'hides' the mobility of the MS from the PDN which 'sees' the GGSN simply as a router with additional log-in functionalities like those of a standard network access server.

Seen from the point of view of an MS, the procedure to establish a connection to the internet consists of the two step:

- first, the attach to the GPRS network which only involves the SGSN (and not the GGSN);
- second, the establishment of a PDP context via the SGSN to a GGSN.

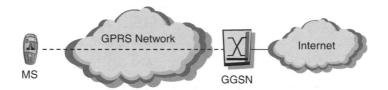

Figure 6.5 Login of an MS into a GGSN

Therefore no mobility management functions are performed by the GGSN. The GGSN, to a certain extent, occupies the same hierarchical level as a gateway MSC in the circuit switched core network that connects the PLMN with another external circuit switched network (Figure 6.5).

The GPRS network is completely transparent as far as the IP packets from the MS are concerned. The first device that is capable of reading the header of the IP packets generated by the MS is the GGSN, which forwards the incoming packets into the connected network based on the destination address.

The following functions are performed by the GGSN:

- packet handling functions (forwarding incoming user data from a SGSN to a Gi interface and vice versa, priority switching in the case of quality of service);
- session management functions (PDP context activation/deactivation to establish/release a connection with a certain SGSN);
- user administration (e.g. IP address and DNS address assignment for the registered MS);
- charging data collection (statistics can be collected about the amount of transferred data to generate file for billing);
- performance management (to measure the traffic within this part of the network);
- fault and maintenance management (to discover problems during data transmission and procedures);
- realizes protocols and interfaces (Gn to connect to SGSN, Gi to connect to PDN); these interfaces are implemented with modules running protocols such as GTP and IP.

Like the SGSN the hardware of a GGSN can be realized with an ATM switch or a router with additional processor platforms, as illustrated in Figure 6.6, which shows the main elements of a GGSN.

The physical realization of the GGSN depends on the manufacturer. Some produce stand-alone units in separate racks while others produce modular GGSNs integrated into the same rack.

An overview of the modular functions is given below:

- *GTP*: as in the SGSN, the GTP protocol is implemented to establish and release a PDP context and forward packet data. Furthermore, IP and UDP are implemented in this module as well. IP is necessary to route the traffic between a GGSN and an SGSN through the Gn Network.
- *ISP*: this module essentially is a router that connects the PDN with the MS. The main functions are the user registration (like IP address administration or data exchange with the RADIUS server) and the forwarding of data.

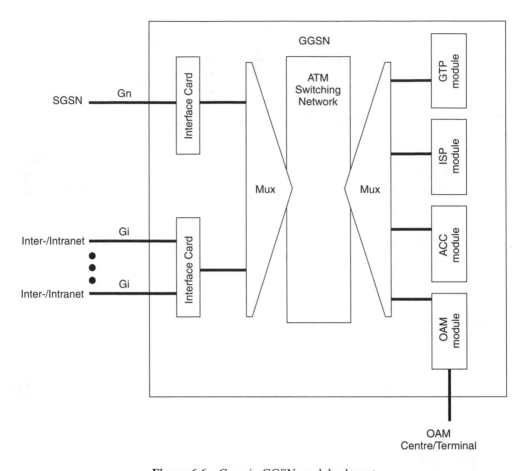

Figure 6.6 Generic GGSN modular layout

- *ACC*: carries out statistical and accounting functions and stores charging data records which can be retrieved by an accounting and billing centre. It is necessary to store billing data on both the SGSN and the GGSN since the former is responsible for network access and the latter is responsible for internet access.
- *OAM*: realizes all operation, administration and maintenance functions to operate the system, e.g. remote configuration/update of GGSN software, alarm notification and fault clearance.
- *Interface cards*: implement the protocols of layers 2 and 1, i.e. Ethernet 100BaseT or ATM/STM-1 for the Gn and the Gi interface.

As with the SGSN, redundancy is essential and can be implemented in different ways. As a minimum redundancy a manufacturer duplicates the entire GGSN hardware and operates the two systems on a load-sharing basis, i.e. both are operational and each can take over the full load if the other experiences problems or fails. The load sharing can easily be implemented since the SGSN contacts a certain GGSN via a DNS request. The DNS returns the two IP addresses of the two parts of the duplicated GGSN. Every part can be contacted individually. Now the load sharing works in such a way that the SGSN contacts one part of the GGSN only every other PDP request. In case of a failure of one part of the GGSN, the SGSN would forward all data only to the working part.

6.3   Access Network PCU – SGSN (Gb Interface)

The access network, which connects the PCU with the SGSN, is standardized down to layer 2. It runs the frame relay protocol on layer 2 (plus some additional network service information) and BSSGP, LLC and SNDCP (only transmission plane) on the higher layers, as can be seen in Figure 6.7. While the BSSGP header is generated by the BSS the headers of the layers above are created by the MS. These headers are transparently transmitted through the BSS.

In terms of GPRS, the network service header (layer 2) contains the header of the switching protocol (FR) plus information about the so-called network service virtual connection. Seen from the point of view of traditional data networks, the additional network service information is not a part of the frame relay header. Therefore, this information is not read by FR switches in the Gb network, which is between the BSS and the SGSN. The switching decision is only based on the FR header.

Since the layer 1 protocol in general is E1, it is possible to establish a variety of physical connections between a PCU and an SGSN. The three possibilities shown in Figure 6.8 are described in the following:

- The most flexible solution interconnects the PCU and SGSN via a pure FR network. This network consists of individual FR switches interconnected via transmission paths. Every switch is able to multiplex virtual channels from individual PCUs into one 'FR resource' consisting of a number of time slots combined into a so-called 'channel group'. More detail about FR switching is given later in the text. Essentially it is possible to multiplex several virtual FR channels into one FR channel group. Within this channel group all connections share the physical bandwidth, which means that a single connection can take the entire bandwidth if all other connections are not transmitting any data. Therefore a very efficient use of bandwidth is possible.

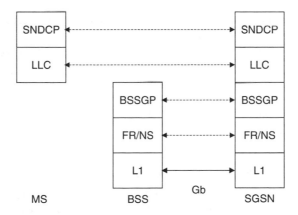

Figure 6.7   Layers of the Gb interface

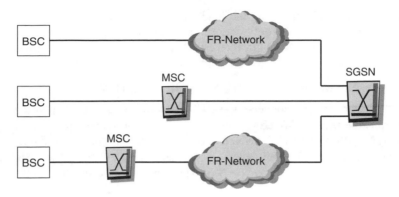

Figure 6.8   Three possibilities to interconnect a PCU with a SGSN

- A second possibility is the interconnection between PCU and SGSN via a leased line consisting of one or several time slots. These time slots are switched purely on layer 1 although the layer 2 protocol is still FR. For this reason it is not possible for more than one PCU to allocate a FR resource (channel group) as seen from the SGSN side. Every time slot of a channel group is switched individually on layer 1 through the network. It is not possible to multiplex several FR connections into one channel group on the way through the network. The bandwidth between one PCU and the SGSN is fixed. A channel group that arrives in the SGSN contains only one connection from one PCU.
- In a third solution, it is possible to mix leased line and FR network with the benefit of an efficient use of bandwidth and a flexible configuration of connections within the FR network.

Although the first solution is the most flexible, only telecommunication companies that install a new Gb network would interconnect PCU and SGSN solely via FR since the investment in such an infrastructure is quite high. Most providers will choose the second or third solution.

Using frame relay it is possible to scale the bandwidth between PCU and SGSN in a flexible way with physical data channels starting at 64 kbit s^{-1}. One or several time slots are combined in a channel group, which is seen as a hardware resource from the FR layer (with a defined physical line rate). FR is now able to multiplex one or more connections into that resource. The individual FR connections are identified via a connection identifier called DLCI (data link connection identifier), which is transported in the header with every frame. The payload of the frames contains the BSSGP packet, which is transparently transported through the FR network from the PCU to the SGSN. All connections in the Gb network (via FR switches and via MSCs) are permanent. The end-to-end connections are configured once to establish communication between the two units.

6.4   Core Network SGSN, GGSN (Gn Interface)

The network interconnecting SGSN and GGSN (in the following text referred to as the 'Gn network') is defined down to layer 3. The protocol stack in Figure 6.9. shows the protocols running on the Gn interface. The network transmits the user data in a GPRS

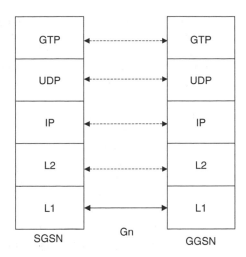

Figure 6.9   Layer model of the Gn interface

tunnelling protocol (GTP) frame which is encapsulated in UDP and IP. IP is responsible for the routing decisions through the Gn network to transmit the GTP frames between an SGSN and a GGSN. The two lowest layers are not specified in GPRS. Any switching technology is possible, although ATM for the wide area connections and Ethernet for the short distances seem to be becoming the favoured technologies.

An example of the configuration of a typical Gn network can be seen in Figure 6.10. The diagram shows the interconnection of two SGSN and two GGSN via an ATM network

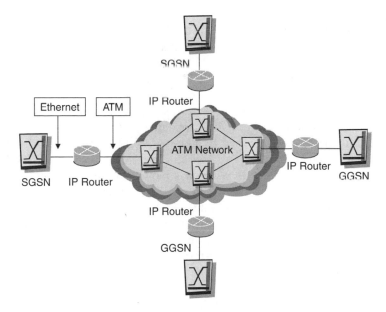

Figure 6.10   Structure of a typical Gn network

with edge routers based on IP. The IP packets from the SGSN/GGSN are transmitted in Ethernet frames to the corresponding edge router of the Gn network. Normally you find the SGSN/GGSN and the router in the same location, both with an Ethernet interface. To transmit the IP packets to another SGSN/GGSN an ATM network is used which is accessed by the router. Therefore the router also holds an ATM interface that is based on E1, E3, STM-1, etc. as layer 1 protocol. The individual routers are interconnected through the ATM network, which switches the data using permanent connections. A separate ATM connection has to be configured between every device around the ATM network to establish communication between these devices.

Seen from the GTP layer, all SGSNs and GGSNs are transparently interconnected through the Gn network since no device – neither the routers nor the switches – read the GTP header. The switching and routing decisions are based only on layer 2 (ATM, Ethernet) in the switches and layer 3 (IP) header in the routers. Since the IP header is only able to identify the hardware unit, additional information is needed to identify the user who transmits the data between SGSN and GGSN. This information is kept in the header of the GTP frame.

Although, at the present stage, the UDP header does not fulfil any purpose (UDP addresses via the port address the application of the higher layer, which at the present stage, is only GTP and therefore the UDP port address of every packet is the same), it is necessary to implement a layer 4 protocol. Since UDP has the smallest overhead, it is the best solution at the moment.

6.5   Additional Elements in the Core Network

Two units in the core network are mandatory to provide mobility management and session management functions such as routing area updates or PDP context activation. These are the DNS Server and the HLR.

While the HLR already exists as a part of the NSS, the DNS is a completely new unit in the core network. Normally the hardware used as a DNS server is a workstation which is directly connected to the Gn network and addressed via an IP address. This address can be found in the configuration of every SGSN. The DNS can be seen as the 'telephone book' of the GSN which can be used by every SGSN to find out the IP address of any other SGSN and GGSN within the network. The software running on this server is based on the DNS protocol from the TCP/IP protocol suite.

One example is the resolution of the IP address of an SGSN. In the case of an inter-SGSN routing area update, a subscriber registers in a new SGSN. This SGSN needs to create an SLR entry for the subscriber who has moved to a new routing area and, with this move, also changed to a new SGSN. In order to contact the old SGSN (which has an existing SLR entry for this subscriber) the new SGSN needs the IP address of the old SGSN. The mobile station does not transmit this address but sends only the location and routing area information of the old SGSN. Therefore a resolution of location and routing area into a SGSN IP address has to be performed, a task performed by the DNS Server. The database of the server holds the IP addresses of all SGSN within the network with a reference to the corresponding location and routing areas that are administered by every individual SGSN. A request with a given location and routing area to the DNS server is answered with the IP address of the SGSN that administers the given area.

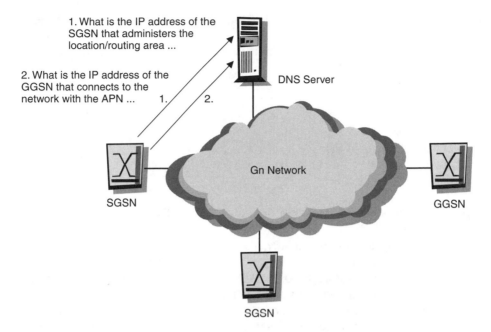

Figure 6.11   Two possible DNS requests from an SGSN

A second task of the DNS server (Figure 6.11) is the resolution of GGSN IP addresses. During the process of a PDP context activation, the mobile station transmits to the SGSN the name of the network it wants to be connected to – this is the so called access point name (APN). The APN is a logical name, in the same way that an Internet address such as *www.tfk.de* is a logical name (for an IP address), with the difference that the APN does not address an application server but a GGSN. Since this APN is the name of a network connected via a particular GGSN, it is necessary for the SGSN to find out the GGSN IP address before it is possible to establish a SGSN-GGSN connection (GTP tunnel). This information is also stored in the database of the DNS server. A request with a given APN is answered with the IP address of the GGSN that connects to the network specified with the APN.

While the DNS server is purely for the packet switched core network, the HLR is a unit that offers services to both the MSC and the SGSN. Therefore the entries in the HLR must be extended with the addition of GPRS specific information.

The GPRS register (GR) is an extension to the HLR that stores the general subscription data allowing a GPRS user to attach to, and make use of, the GPRS PLMN. The GR also contains the following records related to the PDP context(s) of every subscriber:

- *PDP type*: indicates the type of protocol used by the GPRS mobile subscriber for a certain service.
- *PDP address*: a mobile subscriber has one or more temporary and/or permanent network layer addresses. These addresses are activated and deactivated through mobility management (MM) procedures.

- *Quality of service (QoS) profile*: the QoS for a certain PDP context is described with different parameters.
- *Access point name*: the APN field contains either an APN network identifier or the wild-card value.

These subscriber records serve to identify:

- the routing area,
- the network elements of the GPRS subsystem,
- the network elements of the base station subsystem,
- the GPRS mobile subscriber,
- the mobile equipment,
- a packet data network (PDN) access from a GPRS PLMN.

When a subscriber attaches to a SGSN, these data are transmitted from the HLR to the SGSN and are stored as an SLR entry. The connection between an SGSN and the HLR – the Gs Interface – is based on a PCM30 frame. The signalling information is transmitted in time slots between these two units.

With the additional elements HLR and DNS in the Gn network, it is only possible to establish a PDP context between an SGSN and a GGSN in the same PLMN. The establishment of a PDP context between two different PLMNs, e.g. the SGSN and the GGSN are not in the same PLMN, is realized via the Gp interface. Essentially the transmission of data via the Gp interface is based on the same protocols as the Gn interface. Seen from the hardware point of view, there are some differences since the network is now open to other networks and therefore needs security mechanisms to protect the network from unauthorized access. This functionality is implemented in the border gateway which consists of a router with additional features such as a firewall.

In principal, any network element that controls data traffic for security reasons can be called a firewall and as a result the term 'firewall' can in fact refer to a variety of protection strategies. There are three main types of firewall, each using a different approach:

- Packet filtering (screening): based on routers (lower protocol levels);
- Proxy server: high-end proxy server gateways (upper protocol levels);
- Stateful inspection: operates on the bit level in packet headers.

Packet filters (also known as 'screening routers') operate in the lower layers of the network protocol stack and can look at information related to the hard-wired address of a computer, its IP address (network layer), and even the types of connections (transport layer). A set of rules can be implemented to filter packets based on this information. Typically for GPRS, such firewalls will reject packets from certain sources.

Proxy server gateways are application-level devices that can provide more opportunities for monitoring and controlling network access. A gateway firewall acts like a middleman, securely relaying messages between internal and external clients and services. There are two types of proxy servers: circuit-level gateways (offering basic proxy functions) and application-level gateways (offering extensive packet analysis). With proxies, security policies can be much more powerful and flexible because all of the information in packets can be used by administrators to write the rules that determine how packets are handled by the gateway. They are typically used to defend against attacks by hackers.

Stateful inspection techniques avoid one of the main problems with proxies, i.e. they must evaluate a lot of information in a lot of packets, which affects performance and increases costs. With stateful inspection, packets can be screened based on comparison of their header bit patterns. Typically, such firewalls are optimized for GPRS to provide stateful inspection of the key GPRS protocol – GTP (GPRS tunnelling protocol).

6.6   Additional Elements at the Gi Interface

The Gi interface is the gateway from the GPRS core network to a connected PDN such as the Internet. Generally, this point specifies the end of the GPRS network, as the connected PDN belongs to a different network operator. Nevertheless, there are some elements that can be implemented at the Gi interface to obtain additional functionalities.

A RADIUS server, at the Gi interface, supports all features that can also be found in a network access server solution (e.g. Internet access via modem over a telephone network) supported by a RADIUS server. The term 'RADIUS' refers to both the application and the protocol for communication with the RADIUS server. It is taken from standard remote log-in solutions and stands for 'remote authentication dial-in user service'. The RADIUS server holds the user database with log-in names, passwords and user specific profiles. Before a user is allowed to enter the network, these data are checked. A RADIUS server is also capable of collecting statistics for accounting. The GGSN is the hardware unit that connects to the RADIUS server to transmit the user data for verification. In this case the GGSN acts as the RADIUS client as illustrated in Figure 6.12.

Another device at the Gi interface is the DHCP server (dynamic host configuration protocol) – another protocol taken from the IP world. Instead of configuring the IP addresses that can be given to the MS in every GGSN individually, it is possible to establish a central unit that administers the available IP addresses. In the case of a PDP context activation, the GGSN contacts the DHCP server to receive an IP address for the user. The advantage lies in the centralized administration of the address pool.

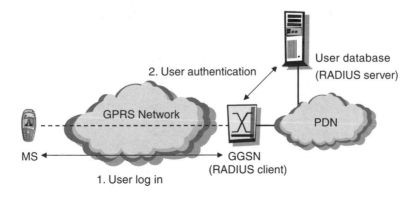

Figure 6.12   User registration and authentication at the RADIUS server via GGSN

6.7   Connections Towards the GSM Network

In addition to the hardware units and networks presented here, there are a number of interfaces towards the 'classical' GSM network, which are running SS7 signalling. The Gr interface, connecting the SGSN to the HLR (as illustrated in Figure 6.13) is standardized in GSM recommendations 09.02 for MAP, 09.16 and 08.06 for all layers including modifications to the message transfer parts (MTP) and the use of E1 and T1 for layer 1. The other SS7 interfaces are between:

- SGSN and MSC,
- GGSN and HLR,
- SGSN and SMS centre.

The Gs interface is running BSSAP+ (a subset of BSSAP see Figure 6.14) to implement so-called 'combined procedures' for mobility management. In the case of a routing area update, it is possible to update the SLR entry in the SGSN and send a location update to the corresponding MSC/VLR via the Gs interface. This procedure reduces the load on the signalling channel of the radio interface, which is normally the bottleneck in the GSM/GPRS infrastructure, and therefore saves resources.

The Gc interface can be realized in two ways, either by a direct link between GGSN and HLR, or via a GTP – MAP inter-working function (IWF) to be implemented at a

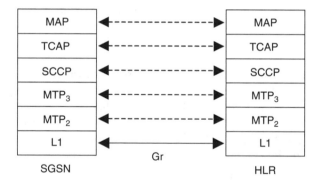

Figure 6.13   Protocol layers of the Gr interface

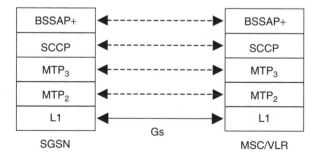

Figure 6.14   Layer model of the Gs interface

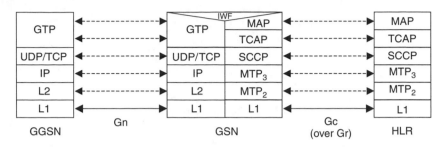

Figure 6.15 Protocol stacks of the Gc interface realized via an IWF

GSN node in the GPRS backbone network (Figure 6.15). In this way, Gc can be realized via Gr from an SGSN. The Gs interface is implemented to support network-requested PDP context activation. In the case of a PDP context activation that is initiated from the PDN connected to the GGSN, the GGSN needs to resolve the address of the SGSN in which the called user is currently registered.

7

Terminal Equipment

7.1 Types of Terminal Equipment

At the introduction of GSM, the terminal equipment was simply a mobile phone. This was something similar to the good old receiver to hold in your hand, or something larger to build into cars, and use for making and receiving voice calls. Things have developed quite significantly since then. Nowadays, mobile phones have built-in organizers and games, along with various databases. They are used to send and receive text messages (GSM uses SMS rather than the Internet text message standard email) and you can even log on to the Internet, e.g. via WAP. However, the fact remains that for quite a considerable number of mobile phone users, the possibility of making and receiving telephone calls while on the move is still all that matters. They do not feel the need to make use of the other capabilities that are present in their mobile. For many such users 'a phone is a phone', and they are not motivated to learn and use WAP, SMS, games, the organizer and services, as these are seen as unnecessary extras. Even for subscribers who regularly make use of SMS and other services, the next jump to mobile internet access and other services is also a question of acceptance.

It is GPRS that might actually change our perception of what a mobile device actually is and what we can do with it.

7.1.1 GPRS Enabled Mobile Phones

'A classic mobile phone with GPRS ability' is probably what most people think of when they hear of GPRS for GSM networks. As GPRS is often mentioned together with 'Internet access', a common misconception is that GPRS is just a newer version of WAP. While it is true that WAP content can be sent via GPRS, users must understand that GPRS is a bearer service that can be used to enable other services (such as those based on WAP), i.e. it can be used to access the Internet either via the WAP gateway or directly via the GGSN as we have seen in the preceding chapters.

To take full advantage of the GPRS functionality, additional features have to be implemented and then again users have to accept them. In this respect, the fact that a mobile device can utilize GPRS is not as important as what GPRS enables the mobile device to do, i.e. what the benefits are for the user. An example here would be mobile devices with

GPRS Networks.   G. Sanders, L. Thorens, M. Reisky, O. Rulik, S. Deylitz
© 2003 John Wiley & Sons, Ltd.   ISBN: 0-470-85317-4

the ability to execute Java Micro Edition programs such as single player games. Such devices do not really need GPRS. However, if the user can be motivated (by marketing, peer pressure, trends, etc) to download new games *frequently*, then GPRS would offer a real benefit for the user. The move to on-line gaming, in which players compete in a virtual environment against other players, will more fully utilize the benefits offered by GPRS, i.e. many small data packets representing game actions and reactions will be sent and received, and the user will pay for the total volume of data rather than the duration of the game. The popularity of such interactive multiplayer games on the Internet has already translated into a huge success for i-mode in the field of mobile gaming.

GSM enjoyed healthy growth from the very beginning, but two applications, namely SMS and prepaid accounts, had extraordinary success and thereby helped GSM to achieve enormous market penetration and acceptance worldwide. These were the so called 'killer application'. At the time of writing, some are claiming that the long sought after 'killer application' for mobile data networks has been found, and again it is a messaging service. Multimedia messaging service (MMS) is built into recent top end mobile phones, enabling them to send and receive messages with attached pictures (taken with built in digital cameras), sounds, or predefined animations. Because of the much larger message size compared with plain text SMS (depending on picture resolution or sound-file length), it is only feasible with high data-rate bearer services (even better with a packet oriented service) like GPRS. Only time will tell whether it actually is a 'killer application'. When this was first claimed, there were not even 100 000 MMS enabled mobile phones in use worldwide, and these were used to send 1 to 5 MMS per week averaging some 30 kbytes each. Compared with SMS traffic and revenue, this is still completely negligible, but at least the fast acceptance gives hope to the service providers that there really is some demand for higher bandwidth data services on the market.

Using the mobile phone to access the Internet (without a laptop or PDA attached) is enabled by the wireless application protocol (WAP). This protocol reduces the load on the air interface, compared with the TCP/IP suite from the classical Internet, and content can be adapted to small mobile phone displays and slow processors, etc. (see details in Chapter 10) For this, the mobile device needs to be equipped with a WAP browser. The connection to WAP compatible content (or any other content in the World Wide Web) can be established using circuit switched data (CSD) transfer (which made WAP very infamous for being slow and expensive), by HSCSD (for which you pay for connection time) or by GPRS. Sending WAP data via GPRS is not only faster by several factors, but also much more resource efficient as no time slots are allocated when data needs to be transferred. Thus access to the Internet via GPRS can be offered to the customer much more cheaply than CSD or even HSCSD, which was unfortunately not the case at the introduction of WAP. A special version of a GPRS mobile phone will be a 3G handset that has to include 2G functionality for compatibility reasons. As long as 3G networks do not offer full coverage (and it is not sure whether they ever will), the 3G devices on the UMTS side will most likely offer both GSM and GPRS capabilities in order to support basic services throughout a national network.

7.1.2   GPRS Enabled Mobile Computers

Another application of mobile phones has been their use as a modem, i.e. just as a normal modem enables a computer to access a data network via telephone, so a GSM mobile can act as a modem for a laptop to access a data network. The computer is connected to the phone via an interface cable (or other connectivity technology such as Infrared or Bluetooth) and uses the mobile phone as a modem to contact an internet service provider (ISP) or company intranet. These circuit switched data (CSD) connections, as the name suggests, are fixed and require that resources (time slots) be allocated to one user for the duration of the session. Further, the original WAP via CSD required the use of a special network element called a remote access server (RAS) to interwork between the GSM CS environment and the IP based Internet environment. As we have seen, browsing the Internet or sending and receiving e-mail with a mobile computer is not a continuous data flow, and thus GPRS is a far more suitable bearer than CSD, and does not require the services of a RAS.

Some users do not want to have to connect and disconnect their mobile phone with their laptop or PDA. Perhaps they want to be connected, with their email server for example, all the time via the 'always on' functionality of GPRS. In this case a GPRS module can be implemented directly in the computers or plugged in semi-permanently with a PCMCIA card. Such a GPRS enabled mobile computer may be connected permanently to the Internet (always on), just as if the user would sit in his office connected via LAN cables. As long as he or she does not actually request any web pages or perform an email check there is no traffic and thus no costs. As soon as there is data to be transmitted, there are of course higher costs than in fixed data networks. This can be thought of as the 'price of mobility'.

Actually, a subscriber is connected to the Internet as soon as he has activated a PDP context. To stay 'always on', the PDP context must stay active. So if the subscriber does not send or receive any data for a while, he will stay on until the PDP timer (around one hour) has expired. Then the PDP context is deactivated and the subscriber is no longer connected to the Internet.

7.1.3 Convergence: GPRS Enabled Smart Phones and PDAs

The increasing convergence between hand-held computing devices and mobile phones, along with the demand for more functionality in fewer devices, is taking us towards a mobile phone merged with a personal digital assistant (PDA). PDA are ever decreasing in size while at the same time offering more and more processing power, more memory, bigger screens and, most important, a lot of versatility. There are thousands of commercial or freeware programs for all kinds of purposes waiting to be installed in order to generate a highly customized miniature computer. Add a GPRS module to one of these PDAs and you will end up with a mobile office or mobile Web surf and e-mail station.

The mobile handset vendors are approaching this from the other side. A GPRS-enabled mobile phone is given a bigger screen and a more powerful processor, and many useful programs are added to the handset's (still mostly proprietary) operating system. The result here is a so called 'smart phone'. A smart phone's operating system is often not as versatile as a PDA's operating system. A GPRS-enabled PDA is often not as small as a smart phone and might lack integration of the GSM/GPRS module, but the difference between a GPRS-enabled PDA and a PDA-enabled mobile phone will eventually disappear, since they are both an expression of the same idea.

The data intensive 'mobile office' parts in particular (of these devices) rely heavily on GPRS. Although there have been HSCSD, and even CSD only, smart phones and PDAs, the real benefit of 'always on' and reading Web content and emails without worrying about the tariff counter in the background, comes only with packet oriented services like GPRS or its designated successor, UMTS.

7.1.4 Other Kinds of Connectivity

An end-device may have additional interfaces for other standards of mobile communication. Three current examples are Bluetooth, Infrared (IR) and wireless LAN (WLAN). For each of these, the idea is the same: offering a direct interface with high transmission rates between two end-devices. These three technologies define wireless connections over short distances (usually up to 50 metres), although it must be said that IR is limited by the necessity for a clear line of sight between the transmitter and receiver.

It could be very useful for a GPRS-enabled mobile device to have such an additional interface. For example, a direct data connection with high throughput is possible between a laptop/PC and the GPRS mobile phone without the need for plugging in cables. Bluetooth is especially useful in this case as it is a radio connection which means that the mobile can be picked up and moved around during voice calls while still 'connected' to the laptop for GPRS transfers (this is not possible with IR). If the mobile GPRS device is also WLAN equipped, the user can take advantage of wireless LAN services (often offered for free) instead of paying for GPRS service. The range of WLAN can be up to 500 m, enabling a data rate of 11 Mbps. So, such an end device may make data transfers via GPRS while mobile or when other access technologies are unavailable, and perform local data exchanges with these other technologies when it is convenient to do so.

It is also planned that transitions similar to handovers should be possible between WLAN and GPRS, for example when an area covered by a pico-cell borders on an area covered by WLAN.

7.2 Multi-slot Classes and GPRS MS Classes

The first section dealt with the GPRS device itself. Here we will look at what the network needs to know about the terminal equipment in order to provide services at an adequate QoS level. It has already been discussed in Chapter 2 which Quality of Service QoS Classes can be requested by a GPRS mobile and granted by the network. However, there are still differences in what each individual terminal equipment can cope with in terms of the maximum number of time slots it can use, and the execution of simultaneous services (data and voice). The former is described by the so called 'multi-slot class' of the terminal equipment and the latter by its 'GPRS class', which defines its mode of operation.

7.2.1 Multi-slot Classes

A multi-slot configuration consists of multiple packet or circuit switched channels together with their associated control channels, all allocated to the same mobile phone. The multi-slot configuration for a single MS can occupy up to 16 physical channels, i.e. up to eight UL time slots and up to eight DL time slots (each defined by their time-slot numbers TNs), which lie on the same uplink/downlink frequency pair.

A mobile may be allocated several UL and DL packet data traffic channels (PDTCH/U or PDTCH/D) and according to the GPRS specification 05.02, the UL should have the same TNs as the DL. For example, this means that if an MS gets four time slots (TS1, TS2, TS3 and TS4) on the downlink frequency and one time slot on the uplink, then the uplink time slot has to be either TS1, TS2 TS3 or TS4 (not TS 5 to 7) on the uplink frequency.

The multi-slot configuration depends on the type of mobile:

- Type 1 mobiles cannot transmit and receive simultaneously. Hence their multi-slot usage is limited by the time required to send the several UL time slots and to receive the several DL time slots.
- Type 2 mobiles can transmit and receive simultaneously and hence can support more time slots.

The descriptions above often cause misunderstanding as it appears that the mobile must transmit and receive simultaneously as the same time slot number is used on the UL and the DL. In actual fact, reception and transmission by the mobile are separated in time, i.e. the mobile first receives a time slot then pauses before transmitting. This is why the first GPRS mobiles can use a maximum of five time slots at once (multi-slot class 12), e.g. 1 UL and 4 DL, or 2UL and 3 DL. The pause between receiving and transmitting is the equivalent of three time slots in duration. Hence, in order to use more than five TS a mobile must be able to send and receive simultaneously.

There are 29 multi-slot classes defined, each with specific parameters. Key parameters for multi-slot configuration are illustrated in Figure 7.1, and described in the following:

- Rx, the highest number of timeslots on which an MS can receive;
- Tx, the highest number of timeslots on which an MS can transmit;
- Sum, the highest total number of timeslots that can be used in the uplink and downlink. For example, the expression $Tx = 3$, $Rx = 3$, $Sum = 4$ refers to multi-slot class 7 and means that a mobile of this class can use one of the following combinations of time slots:

$$3 \text{ DL} \quad \text{and} \quad 1 \text{ UL} \quad 2 \text{ DL} \quad \text{and} \quad 2 \text{ UL} \quad 1 \text{ DL} \quad \text{and} \quad 3 \text{ UL}$$

In any case, the total number of timeslots cannot be greater than four. This parameter 'sum' is only applicable to those multi-slot classes which describe mobiles that cannot send and receive simultaneously.

7.2.2 GPRS MS Classes

As mentioned in Chapter 2 there are three different classes of GPRS terminal equipment. The GPRS-enabled mobile device should be able to handle both voice and data traffic, and the class describes the mobile's capabilities in this respect, as in the following sections.

7.2.2.1 Class A

Class A 'Classical'

The MS is attached to both GPRS and other GSM services. The MS supports the following for GSM and GPRS simultaneously:

- attach,
- activation,

Multislot Class	Rx	Tx	Sum	Type
1	1	1	2	1
2	2	1	3	1
3	2	2	3	1
4	3	1	4	1
5	2	2	4	1
6	3	2	4	1
7	3	3	4	1
8	4	1	5	1
9	3	2	5	1
10	4	2	5	1
11	4	3	5	1
12	4	4	5	1
13	3	3	NA	2
14	4	4	NA	2
15	5	5	NA	2
16	6	6	NA	2
17	7	7	NA	2
18	8	8	NA	2
19	6	2	NA	1
20	6	3	NA	1
21	6	4	NA	1
22	6	4	NA	1
23	6	6	NA	1
24	8	2	NA	1
25	8	3	NA	1
26	8	4	NA	1
27	8	4	NA	1
28	8	6	NA	1
28	8	8	NA	1

Figure 7.1 Multi-slot classes

- monitoring,
- invocation,
- traffic,
- mobility management.

'Simultaneously' means that the mobile is required to support both GPRS packet switched services and GSM circuit switched services (including SMS) at the same time.

The mobile user can make and/or receive calls on the two services simultaneously, subject to the QoS requirements. A minimum of one time slot will be available for each type of service (circuit switched and GPRS) when required (Figure 7.2).

The engineering effort needed to realize the Class A requirements is very high as the mobile device has to monitor both GPRS and GSM signalling channels for incoming

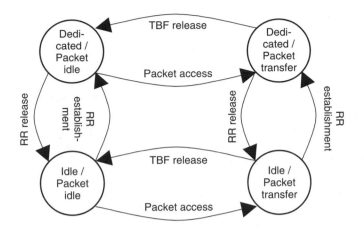

Figure 7.2 Modes of operation and their transitions for Class A 'classical'

calls (paging) and other messages (if no common mobility management is established). It also has to support two different frequencies for voice and data services at the same time while still monitoring neighbouring cells for reception level and broadcasts. The solution to this problem is described in Release 99, *GSM Spec. 03.55 'Dual Transfer Mode'*.

Class A Dual Transfer Mode DTM

The basic idea of a dual transfer mode (Figure 7.3) device is that packet switched data can be sent:

- on a time slot that is currently being used for a circuit switched connection;
- on other time slots on the same frequency as the current CS connection.

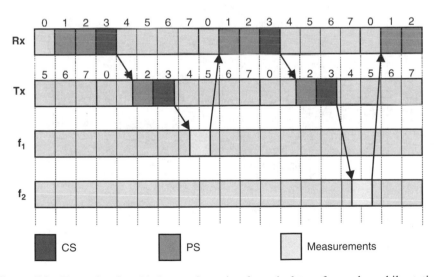

Figure 7.3 Example of multi-slot configuration for a dual transfer mode mobile station

There are two restrictions for the multi-slot configuration of a DTM device:

- only one time slot can be allocated to the circuit switched connection,
- additional time slots for packet switched use have to be neighbouring ones.

There are some minor changes on the A interface and to signalling evaluation, but overall the impact of the dual transfer mode on the core network is negligible.

As described in Chapter 2, the behaviour of a DTM device is more like the Class B mode of operation. The DTM device can change from idle/packet idle to either packet transfer (GPRS data) or dedicated (GSM voice) mode and back, exactly like the Class B device. However, in addition it can change from dedicated mode to dual transfer mode, i.e. from sending only voice, to sending data on the CS time slot or on additional PS time slots on the same frequency). Since this happens without changing the radio frequency and without leaving the dedicated mode (the 'call') first, this mode is easier to implement into terminal equipment. The possibility of running voice and data services simultaneously makes it a real Class A mode of operation (Figure 7.4). It should be noted that DTM can only be initiated on a CS channel, an established packet channel can not change to DTM.

7.2.2.2 Class B

A mobile station in Class B mode of operation, is attached to both GPRS and other GSM services, but the MS can only operate one set of services at a time, i.e. the MS is either in dedicated mode or in packet transfer mode but not both. Three cases must be considered:

- When the MS is in both idle mode and packet idle mode, it should be able to monitor paging channels for both circuit switched and packet switched services depending on the network operation mode (NOM I, NOM II or NOM III). With NOM III, the network

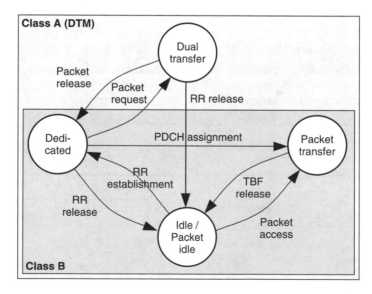

Figure 7.4 Modes of operation and their transitions for Class A 'dual transfer mode' and Class B

performs the paging for circuit switched and packet switched services on different paging channels. In this case, an MS should either attempt to listen to both paging channels with priority for the circuit switched service or revert to Class C mode of operation. With NOM I or NOM II, the network performs both the paging for circuit switched and packet switched services on the same paging channel, so the mobile station should respond to paging messages for both services.

- There is no requirement for the MS to monitor the packet paging channel when in dedicated mode, however it can move from dedicated mode to packet transfer mode, but not back.
- When an MS is engaged in packet data transfer, it may receive paging messages for circuit switched calls. With NOM I, the MS receives such a paging message via the packet data channel without degradation of the packet data transfer. With NOM II or NOM III, such paging is not done via the packet data channel but via the GSM paging channel (PCH) or the GPRS packet paging channel (PPCH). When responding to a paging message for GSM services, the MS shall establish the connection for that incoming service (i.e. enter in dedicated mode via idle mode) and suspend GPRS activity. In general, if the user is paged for an incoming circuit switched call during a GPRS connection, the end device shall indicate the presence of the call to the user or the user's application. It shall be possible for the user (or the user's application) to decide how to proceed with an incoming call (e.g. accept the call, indicate that the user is busy, or invoke a call forwarding). Furthermore users should be aware that monitoring paging (in NOM II or NOM III), responding to paging, alerting of circuit switched service, or acceptance or establishment of a circuit switched call during an active GPRS connection may degrade the performance of the established GPRS connection. In some cases, this may cause a failure in an application using the GPRS connection, e.g. a file transfer might be aborted due to a time-out of the application protocol.

7.2.2.3 Class C

The Class C device is attached to either GPRS or other GSM services. If both services (GPRS and circuit switched) are supported, then a Class C device can make and/or receive calls only from the selected (manually or by default) service, i.e. either GPRS or circuit switched service but not both. The status of the service that has not been selected is detached, i.e. is registered in the network as 'not reachable'. The capability for GPRS attached Class C MSs to receive and transmit SMS messages is optional.

The network must support SMS message reception and transmission for GPRS attached Class C MSs (Figure 7.5).

7.2.2.4 Additional Remarks about the GPRS MS Classes

The definition of different GPRS MS classes enables the different needs of the various market segments to be satisfied by a number of MS types with distinct capabilities as described in the previous sections. For example, a PCMCIA card for portable computers may not need a voice capability at all, thus Class C is sufficient. On the other hand, a modern PDA or smart phone needs to support both voice and data, but not necessarily at the same time. The customers should have the freedom to chose a device that best suits their own personal needs. Most current devices are Class B capable, because it is fairly

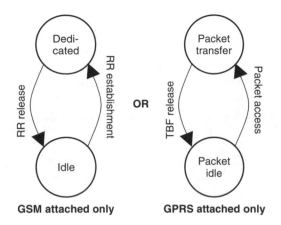

Figure 7.5 Modes of operation and their transitions for Class C

easy to develop and easy to use, but since the introduction of dual transfer mode there is a chance for Class A services to be pushed into the market.

It should be noted, that an MS may be reconfigured, e.g. a Class A MS configured with one time slot for circuit switched plus one time slot for GPRS may be reconfigured as a Class C MS configured with no time slot for circuit switched plus two time slots for GPRS.

7.3   The Settings in a GPRS-enabled Mobile Device

We have now seen all the network elements in a GPRS/GSM system and we know what the mobile devices are capable of. So far this has been quite theoretical. There is still the question of the practical set-up of a GPRS mobile phone and what the settings mean. It will not be possible to give settings for all the different networks in the world, but given the right settings from the operator it should not be difficult to connect a GPRS phone to WAP services or use it with a laptop to access the Internet.

7.3.1   WAP Settings

The first parameter we will look at when entering the Internet profile settings are the WAP settings. The WAP settings are necessary for accessing the WAP gateway and as such are independent of whether classical WAP or WAP over GPRS is used.

There are usually two IP addresses, their respective port number, and a start page. The IP addresses are related to the WAP gateway which serves as an interworking function between the TCP/IP protocol suite of the Internet and the WAP protocol suite of the mobile networks (see Chapter 10), and as a proxy to access the Internet. The second IP address is only used when a PLMN contains more than one WAP gateway. It may be used when the first gateway is busy or out of service.

The port number refers to the security and reliability settings in the WAP protocol stack. The standard setting is 9201 which refers to the use of the whole WAP protocol stack apart from the wireless transfer layer security (WTLS). When the user wants to run security sensitive applications like Internet banking he should switch to port 9203. If the port is switched to port 9200, then WTLS as well as the WTP flow control layer (wireless transport protocol for reliable transmission) are switched off. This is only applicable when applications are either completely error insensitive or retransmission could not be performed anyway as with real time applications, e.g. video or voice streaming. The missing setting 9202 is never applied, because an unreliable but secure connection is not of much use.

Last but not least, the start page refers to the URL that will be loaded at log-in. This could be the default portal of the network provider or a third party. Since this will be loaded every time the user logs on the network, it should not refer to a 'big' page with a lot of kilobytes to download.

7.3.2 CSD Settings

This section is only included for completeness as CSD can be used for data transfer if GPRS is not available.

First there is a CSD dial-up number. This is the telephone number of a modem which enables a circuit switched connection to the WAP gateway. This modem is called the remote access server (RAS) or network access server (NAS), as it implements several hundred dial-up modems for CSD connections.

The call type should be ISDN in all GSM networks, because they are fully digital, but if a RAS in an external analog network needs to be addressed, it can be set to analog as well.

The last settings here are usually the provider specific user log-in and password (e.g. t-d1 and td1). It could also be a user specific log-in and password, but since the check as to whether a user is allowed to use certain services is done in the remote authentication dial in user server (RADIUS) anyway, there is no need to confuse users with personalized settings (this way providers can preset these parameters or configure the mobile stations on air).

7.3.3 GPRS Settings

In order for an MS to be able to access an external data network, it needs the corresponding access point name (APN, see Chapters 4 and 6) of the requested network, e.g. intranet.tfk.de. In the GPRS register (GR – the extension to the HLR), several PDP contexts may be configured for an MS, so that this MS may be able to reach several data networks (i.e. several APNs) each with a different quality of service (QoS). It is also possible that several APNs are defined for the same data network but for diverse services with different QoS. The APNs of the different PDP contexts that are configured in the GR are also listed on the SIM card and they actually are the GPRS settings of the subscriber.

It is not possible to describe the set-up procedure for every GPRS device. However, as a minimum, the user will have to configure the device to use a specific provider using the provider's APN, and set up the connectivity if an external device such as a laptop is to be used. In the case of a laptop, the mobile must be configured as a modem in the laptop's system settings. To do this the user must first decide whether to use the Infrared, Bluetooth or cable interface. This is relatively simple to set up in the mobile as the settings menu simply allows you to select the type of interface to use. The laptop however is more complicated as each interface has its own port number (usually COM1 for cable and COM4 or 5 for IR). It is also possible/necessary to set the bit rate and other parameters used by the selected port. If manual configuration presents difficulties, software for automatically configuring the laptop is usually provided along with the phone on a CD-ROM, or can be downloaded from the supplier's web site.

8

Planning and Dimensioning

8.1 Introduction

The process of planning and dimensioning a network is usually carried out for the following reasons:

- in order to estimate the overall equipment requirements and its deployment;
- to ensure a certain quality of service;
- in order to estimate the overall cost.

Manufacturers of network solutions will perform the above in order to provide a potential customer (network operator) with a price estimate. Hence, the accuracy of the estimate is extremely important as it must be both realistic and competitive, i.e. it must meet the customer's quality of service (QoS) requirements with a minimum of investment.

Dimensioning a network for GSM speech is based on the number of 'busy hour' call attempts which will be made by a given population and the level of 'blocking' that is considered to be acceptable by the operator.

The 'busy hour' is the time when the network requires the highest capacity, i.e. the hour when the most call attempts are made. The volume of calls that cannot be handled by the network (especially during the busy hour) is called 'rejected' traffic. The ratio of rejected traffic to offered traffic gives a factor known as 'blocking'. Acceptable blocking values are typically less than 1 % under normal load, and up to 3 % (maximum 5 %) under very high load.

The requirements of the radio and core networks can only be calculated with respect to the behaviour of a given population of customers. The information regarding such customer behaviour can be based on:

- measurements made in existing networks,
- the customer's (network operator) QoS requirements,
- assumptions based on expected behaviour.

Dimensioning a network for GPRS is somewhat more complicated as we are no longer dealing with individually switched connections and average connection times. Packet data transfer is typically 'bursty' and can range from a few hundred bits to several megabytes.

GPRS Networks.   G. Sanders, L. Thorens, M. Reisky, O. Rulik, S. Deylitz
© 2003 John Wiley & Sons, Ltd.   ISBN: 0-470-85317-4

Added to this is the fact that GPRS transmissions can use one or several time slots, the number of which can either be fixed or vary during the transmission. A further complication is that GSM speech calls can be given higher or equal priority to GPRS, which means time slots might only be available for GPRS if they are not needed for GSM calls. Depending on the network operator, it could be that certain time slots are reserved for GPRS and others for GSM, or it could be that a GPRS transfer starts with four time slots, but due to GSM demand the number of time slots in use varies as they are reallocated for speech calls (if GSM has priority over GPRS).

The following sections look first at the dimensioning of a GSM network, to give the reader an appreciation of some of the difficulties involved, before considering GPRS networks and the need to apply capacity planning concepts to packet data traffic.

8.2 Network Dimensioning

8.2.1 GSM – Circuit Switched Voice

As network dimensioning is dependent on the 'busy hour', which in turn is a reflection of subscriber behaviour, it should be clear that it is not possible to use the same parameters for dimensioning all areas of a national network. The busy hour is not a constant for a network, but changes regionally according to the local traffic patterns and can be influenced by many factors, as the following section will show. In dimensioning a network, the operator is trying to plan a service that meets certain requirements for coverage, capacity and quality, while at the same time minimizing the initial investment and overhead.

Dimensioning requires a great deal of input – some of which is not known and must be estimated – including the geographical area to be covered, the population distribution across the area, the anticipated (regional) traffic volume, anticipated subscriber behaviour, available radio resources (carriers and time slots), acceptable blocking levels, etc. As an initial step towards simplifying the process, the geographical area is subdivided into regions which have relatively uniform internal characteristics, i.e. such a region should not contain zones with significantly varying radio propagation characteristics, and the traffic distribution should be as flat as possible. Areas should be divided and subdivided until these two assumptions are valid in every region – if they are not, this is an indication that a region should be further subdivided.

The radio access network must be planned first as the dimensioning of the core network will depend on the traffic requirements of the base station system (BSS). To this end, we can think of the dimensioning process as a means of defining the number of physical channels (time slots) required for each cell (or sector for subdivided cells) served by a base station (BTS). In other words, the number of time slots determines the number of carriers the BTS needs, along with the signalling channel requirements. This, in turn, determines the requirements for the base station controllers (BSC) and the trans-coding equipment (TCE) – all of which ultimately provides a basis for planning the network switching subsystem (NSS).

As mentioned earlier, there are three critical factors involved in dimensioning the BSS – coverage, capacity and quality – any of which can be the limiting factor that governs the resources planned for each cell/sector. Explained simply, the operator will calculate:

- the number of BTS sites necessary to cover a given area, and
- the number of BTS sites necessary for the anticipated traffic in the area.

The determining factor is the one which returns the highest number of sites, i.e. it is possible to have 'coverage limited networks' or 'capacity limited networks'. *Coverage* limited areas are generally those with low/dispersed population density lying between large population centres. They tend to consist of large cells (up to 35 km) requiring few carriers to meet anticipated demand. *Capacity* limited areas are to be found in large population centres (urban/suburban) with high population densities. They tend to require very small cells (typical radius <500 m) with many carriers. The third factor 'quality' refers to interference, i.e. the ratio of signal to noise (C/I), and depends on frequency reuse and the local environment. This factor is more important when considering data transfer. 'Interference limited networks' will be discussed in Section 8.4.2.

When considering traffic volume and cell/network capacity, it is necessary to think in terms of 'hours of channel usage per hour' – a concept measured in units of 'erlang'. For example, one carrier frequency is divided into eight time slots (channels), so one carrier can carry 8 hours of traffic (channel usage) per hour, i.e. 8 erlang (8 Erl). However, for this to be true, two conditions must be met:

- every time slot must be available for voice traffic;
- every finished call must be instantly followed by a new call.

Clearly, for planning purposes this is unrealistic. Some time slots are always reserved for signalling so the first condition is not always true. More importantly, to fulfil the second condition we must constantly have subscribers dialling or waiting for call completion – this would require an unacceptably high blocking rate. As a result, actual traffic models must consider both customer behaviour and quality of service, i.e. the actual blocking rate must be kept to a minimum.

To calculate the actual capacity of a BTS for a sector or omnicell (or any network element) the Erlang B formula is used. The formula is based on the concept that there is a relationship between blocking (B), the number of available channels (N) and the offered traffic (A).

$$B = f(A, N) = \frac{\dfrac{A^N}{N!}}{\displaystyle\sum_{k=0}^{N} \dfrac{A^k}{k!}}$$

Although telephone calls are random events for the individual, in any given time interval a certain percentage of a population will be making a phone call, and the length of these calls can be averaged. The time interval which directly concerns the dimensioning process is the peak traffic period, i.e. the busy hour. This gives us two important values:

- the number of 'busy hour' call attempts (peak traffic period);
- the mean holding time (average duration).

For example: if it is known (from statistics gathered from existing networks or behaviour modelled on previous networks) that the average subscriber makes 1.2 calls per hour (during the busy hour), and the average holding time is 50 seconds, we can calculate the busy hour capacity requirement of the 'average person' in erlang.

So,

$$50 \text{ seconds} = 0.833 \text{ minutes} = 0.0139 \text{ hours}$$

Therefore,

$$1.20 \times 0.0139 = 0.0167 \text{ Erl per person} = 16.7 \text{ mErl}$$

Such values are commonly expressed in millierlang (mErl) per person

The result of the above calculation will be different for almost every network as customer behaviour varies. The calculation can be generally expressed as the following equation:

$$A = \frac{BHCA \times t_{\mathrm{m}}}{3600} \quad \text{where} \quad A \quad = \text{required capacity per subscriber}$$

$$BHCA \quad = \text{busy hour call attempts per subscriber}$$

$$t_{\mathrm{m}} \quad = \text{average call duration per subscriber}$$

We now need an estimate for the number of subscribers which will be served by each cell in the region. If the average number of subscribers per cell is 1000, we will need enough carriers to provide a capacity of $0.0167 \times 1000 = 16.7 \text{ Erl}$

The offered traffic can be calculated using the Erlang B formula by specifying the number of channels and an acceptable blocking factor. A range of values calculated in this way can be compared with subscriber behaviour to determine the capacity requirements of a given sector or omnicell. Realistic values for cell capacity are given in Table 8.1.

Taking the capacity requirements from the example (16.7 erlang), the number of necessary carriers can be determined by referring to Table 8.1. For the given capacity, one of the following will be required:

- an omnicell with four carriers and 1 % blocking;
- an omnicell with three carriers and 5 % blocking;
- a two sector cell with two carriers per sector and 2 % blocking.

Table 8.1 Relationship between carriers, time-slots, blocking and erlang

Number of carriers	Number of time slots	Signalling time slots	Traffic time slots	Offered traffic (erlang)		
				1 % blocking	2 % blocking	5 % blocking
1	8	1	7	2.5	2.9	3.7
2	16	1	15	8.1	9.0	10.6
3	24	2	22	13.7	14.9	17.1
4	32	2	30	20.3	21.9	24.8
5	40	2	38	27.3	29.2	32.6
6	48	3	45	33.4	35.6	39.5
7	56	3	53	40.6	43.1	47.5
8	64	3	61	47.9	50.6	55.6
9	72	3	69	55.2	58.2	63.7
10	80	4	76	61.7	64.9	70.8

The first and third options both result in over-dimensioned networks, while the second option results in a higher blocking rate. We can see from this simple example that network dimensioning is a compromise between efficiency and quality of service, i.e. allowing higher blocking rates results in a more efficient use of available resources (the highest Erl with the fewest time slots), but at the same time lowers the quality of service experienced by the subscriber. In this case, a new network will probably expect significant subscriber growth, in which case the first option is the most attractive.

The need to balance *capacity* with *quality* is further complicated by the need for minimum frequency reuse distances. The general rule is that if a certain frequency is used in a cell of radius r, then the planner cannot use this frequency again within a radius of roughly $4 \times r$. This minimum reuse distance helps to minimize the interference caused by identical frequencies interacting with each other. There are various planning models in use that first define a group of cells with no reused frequencies, and then repeat this pattern of cells (a cluster) to cover a given area with a minimum of frequencies. Typical models use seven, nine and twelve cell clusters. A four-cell cluster has been introduced which requires frequency hopping to be synchronized between base stations in order to reduce interference. An even more recent development is 'super reuse' which, in effect, places smaller cells within larger cells. This enables shorter reuse distances between the smaller cells. The extra capacity is utilized via handovers between the small cells and the large ones. This can be further enhanced by sectorizing the cells (Figures 8.1 to 8.3).

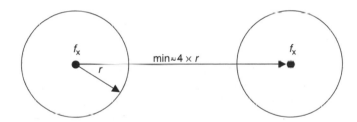

Figure 8.1   Minimum frequency reuse distance

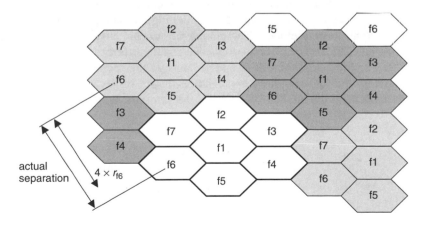

Figure 8.2   Cluster model ensures minimum frequency reuse distance

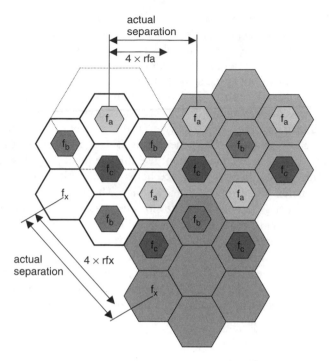

Figure 8.3 Super reuse

The results of the capacity planning stage will allow the number of base station controllers (BSC) to be determined. Here, geographical distribution will also play a roll in determining whether to implement a small number of high capacity BSCs or a larger number of lower capacity BSCs. Ultimately this will also help to determine the number and size of mobile services switching centres (MSC) required. Although here again, anticipated traffic to and from the BSS is not the only consideration – planning must also account for use of services, signalling overhead and protection in case of trunk failure.

The above is intended only as an introduction to the complex world of network planning and for this reason such things as propagation modelling, probability of coverage, antenna height, etc., have not been included. The complexity of the situations faced by network planners necessitates the use of planning tools that enable different parameters to be tested in order to find optimum configurations and are outside the scope of this text. However, the reader should now have enough appreciation of the basics of dimensioning GSM networks to understand the challenges of doing the same for GPRS networks.

8.2.2 GPRS – Packet Oriented Data

Until GPRS can generate similar revenues for the operator as GSM, voice traffic will always have priority over data traffic. For this reason it is possible to implement GPRS in an existing PLMN *without* any extra capacity planning. Low blocking values combined with over-dimensioning (a common result of capacity limited networks), should enable an acceptable quality of service for the low initial volume of GPRS traffic at all but peak traffic times.

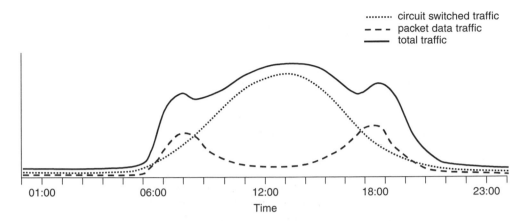

Figure 8.4   Sketch of a hypothetical traffic distribution

Figure 8.4 shows a hypothetical traffic distribution for a commercial city. The voice traffic curve peaks early in the afternoon. However, the peaks in the GPRS curve indicate mobile data transfer from users commuting to and from the office. The third curve shows the total traffic.

In this example we can see there is little conflict between GPRS and GSM traffic. Mobile data users might experience a slightly lower quality of service during the peak traffic hours (12:30 to 14:30) but in general, this network will have sufficient capacity for both GSM and GPRS traffic.

Studies indicate that by the year 2010, the amount of PLMN network resources used for data traffic will have grown to such an extent that it will equal that required for voice traffic. Due to the rise in popularity of the Internet, this event has already happened in the fixed telephone networks of several developed countries. Long before this point is reached in mobile networks, it will be necessary to upgrade existing capacity or, for new networks, to account for the anticipated volume in the planning stage.

There are a number of problems with dimensioning a network for connectionless data traffic. The previous section dealt with circuit switched GSM networks, i.e. networks in which fixed end-to-end connections are set up. The need for one-to-one mapping of time slots between the BSS and the NSS makes it clear that the dimensioning of the NSS depends (to a large extent) on the capacity requirements of the BSS. GPRS networks, however, require a different approach to dimensioning because the transfer of data depends on packet routing, i.e. packets contain addressing information that enables them to be delivered to the correct destination without first setting up a connection (for more information on IP addressing see Chapter 9).

In any packet data network the speed at which a stream of packets can be routed to the destination can be influenced by many factors:

- the maximum rate at which a node can accept data packets,
- the size of the buffer available at a node for one data stream,
- protocol conversion,
- protocol overhead,

- the reliability of the transfer,
- the actual node/network load,
- overload, e.g. due to link failure.

This is further complicated in GPRS networks by:

- the variability of the air interface (CS1 to CS4);
- the variability of available resources (GSM CS calls can have priority);
- use of the GPRS core network for UMTS traffic.

While it is true that many GPRS networks will be implemented with no capacity planning (how this is possible will be explained later in the text), the expansion of these networks will necessitate both long term and short term planning, which must eventually take all the above factors into consideration.

8.3   GPRS Radio Subsystem

On the physical level, it really doesn't matter whether a time slot contains voice or data – the time slot is simply a radio burst transmitting a stream of modulated bits between two stations. In terms of capacity planning, however, the difference is enormous. Voice capacity is calculated in terms of erlang, but the most convenient unit for data is kilobits per second (kbps). This means that if conventional planning techniques are to be applied, the kbps value needs to be expressed in erlang. It is also possible to plan the network in terms of packet transfer (this will be discussed later), and techniques for dimensioning mixed voice/data networks are also under development). As previously described, GPRS has four new coding schemes (CS1 to 4) for the air interface. These values however represent only the transmission rate and do not consider the effect of retransmission. Every packet that is lost or corrupted must be sent again, which has the overall effect of reducing the net transmission rate (throughput), i.e. the perceived bit rate per time slot. As a result, the only realistic measure of throughput is an average. This average throughput per time slot can then be converted to erlang.

The quality of the air interface is the main factor affecting throughput. Quality is defined as the ratio of signal to noise (C/I). Research carried out by ETSI (Recommendation 5.50, Annex M) to investigate the relationship between throughput and C/I is illustrated in Figure 8.5. The two sets of curves, A to D and a to d, reflect the throughput both with and without the use of frequency hopping. This not only illustrates the effect of interference but also demonstrates the value of frequency hopping in improving throughput by reducing the impact of the interference.

Initially only CS1 and CS2 will be implemented so the average throughput will vary between zero and 13.4 kbps according to changes in C/I. In urban areas (where most mobile data users will be), C/I is determined by:

- the level of frequency reuse,
- external sources of electromagnetic radiation,
- environmental effects (such as multipath propagation).

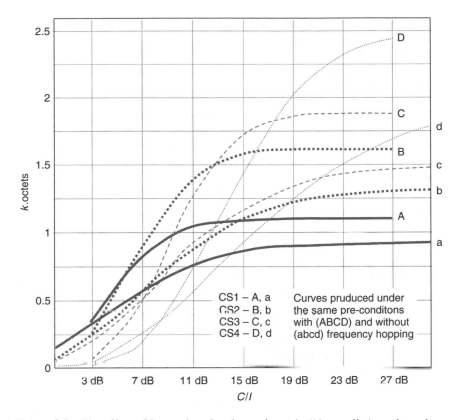

Figure 8.5 The effect of increasing signal-to-noise ratio (Um quality) on throughput

When network planning (capacity and cell size) is constrained by such factors, the result is known as an 'interference limited network'. During off-peak traffic hours there will be less interference, resulting in a better C/I, which in turn yields higher throughput. The opposite is true for peak traffic hours in which the C/I is at its worst.

From the above it is clear that in practice the theoretical maximum throughput (13.4 kbps) is not achievable. Throughput is further reduced during peak traffic hours, when network congestion results in time slots reserved for GPRS being reallocated for GSM voice calls. This effect is especially noticeable when several users must share the same time slots, i.e. there are fewer resources available, and those resources must be shared.

Initially, network congestion will not significantly affect the average throughput which will be dependent mainly on C/I values, which can be measured. As demand for GPRS grows, congestion will start to affect performance and extra capacity will be needed. Here, demand curves should allow adequate estimates to be made.

In Figure 8.5, a C/I value of 11 dB corresponds to a throughput of around 1.39 kilooctets per second (11.2 kbps) using CS2 with frequency hopping and one interfering frequency. Measurements and simulations can be made to confirm or adjust this figure for the actual network being planned. It should be noted that this average throughput refers to the time slot and *not* to the data rate experienced by the user. The data rate at the application (user) level can be lower due, for example, to the effect of retransmissions.

The rate will be potentially much lower if statistical multiplexing is being used to share the time slot between several users (see Section 8.4.3.3) via the MAC header.

Once the average throughput per time slot has been defined in kbps, the corresponding erlang value can be obtained by dividing the total GPRS traffic load (kbps) by the average throughput per time slot (kbps). This will give the total channel usage per unit of time, i.e. 'carried traffic' in erlang. This value can then be used to determine the number of carrier frequencies required. Initially only one carrier will be allocated to GPRS per BTS, and this carrier will either be reserved exclusively for GPRS or shared with GSM voice traffic. This single carrier will be the same one used for the BCCH (broadcast channel) so if it is reserved for GPRS, then a maximum of seven time slots will be available for data transfer. Therefore we have 7 erlang available. This will be sufficient until the total traffic exceeds 7 erlang.

If the single GPRS carrier is shared with GSM voice traffic, the maximum available capacity will vary between 7 erlang and 4.5 erlang. This is because during off-peak hours this carrier will not be used by voice traffic, but during peak hours voice has priority and will need 2.5 erlang (with 1 % blocking, see Table 8.2). Hence, $7 - 2.5 = 4.5$ erlang will be available, on average, for GPRS. So if the erlang calculation based on the average throughput does not exceed this amount, one carrier will be sufficient. The above ignores such effects as 'blocking' and 'queueing' for the packet data traffic – these will be addressed in following sections. The above also ignores the fact that if a cell has many carriers, then the carrier with the BCCH may have more than one time slot reserved for signalling, i.e. it can be that TS0, TS2 and TS4 are reserved for signalling, leaving only five time slots for GPRS.

In early releases, GPRS will use the available time slots on the frequency which carries the BCCH. Later, flexible allocation of all available time slots will increase throughput but with a lower busy-hour GPRS capacity limit (per channel) for the reason given above. Table 8.2 shows the relationship between GSM voice traffic and the capacity available for GPRS, as more carriers are added. Figure 8.6 shows the available 'busy hour' GPRS

Table 8.2 The relation ship between carriers, time slots, voice traffic and GPRS

Number of carriers	Number of time slots	Signalling time slots	Traffic time slots	Offered voice traffic (erlang) 1 % blocking	Capacity available for GPRS (erlang)
1	8	1	7	2.5	4,5
2	16	1	15	8.1	6,9
3	24	2	22	13.7	8,3
4	32	2	30	20.3	9,7
5	40	2	38	27.3	10,7
6	48	3	45	33.4	11,6
7	56	3	53	40.6	12,4
8	64	3	61	47.9	13,1
9	72	3	69	55.2	13,8
10	80	4	76	61.7	14,3

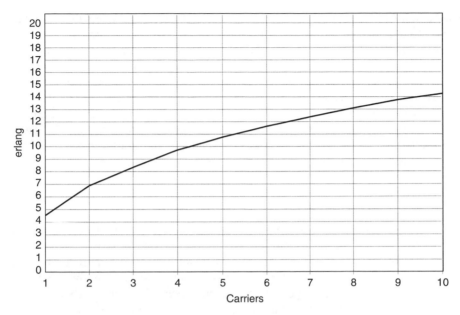

Figure 8.6 Relationship between number of carriers and available GPRS capacity

capacity in erlang as more carriers are added (assuming voice calls have priority and disregarding GPRS 'blocking').

Table 8.2 clearly shows that the resource utilization (efficiency) by GSM voice traffic increases as more carriers are added, i.e. more offered traffic per carrier for the same blocking value. So although the available GPRS capacity is also increasing, Figure 8.6 shows that the increase has an inverse relationship with the increase in voice traffic, i.e. the more carriers are added, the smaller the increase in GPRS capacity, and the flatter the curve.

The above shows that for any given number of carriers there exists an associated amount of capacity available for GPRS. As a result, it is possible to implement GPRS with no additional resources and with little or no extra planning for the radio side. The total available GPRS capacity on the air interface in a given area can be estimated from the average number of carriers per sector cell in the area, e.g. an average of six carriers per sector would indicate an average GPRS capacity of 11.6 erlang per sector. Calculating the exact GPRS capacity for the area would require that GPRS capacity be calculated for each sector (as per Table 8.2) of each BTS and then summing the results to give a total.

The throughput requirements of the PCU will depend directly on the sum of the capacities of the cells (sectors) served by BTSs attached to the BSC, plus the overhead incurred when transmitting the packet data to the SGSN.

The overhead is the additional header and checksum data added when the LLC frames from the mobile are prepared to be sent to the SGSN. In the PCU, a header is added to the LLC frame for the BSSGP protocol and this packet is then prepared for transmission via frame relay (FR) to the SGSN. As the LLC frames have a maximum length of 1543 octets, and the BSSGP header has 53 octets, and a single LLC-BSSGP datagram will always fit into one frame relay frame as the maximum FR payload is 1600 octets.

The amount of equipment required for the PCU will depend on the manufacturer, as throughput limits will vary according to hardware type and software version. PCU installations will almost always be over-dimensioned as the required capacity will rarely be a multiple of the capacity of a single module, i.e. PCU capacity will increase in large steps as modules are added to provide additional throughput, whereas traffic will increase gradually to fill the available capacity. The sum of the PCU throughput requirements will form the basis of the dimensioning of the SGSN to which they are attached. Care must be taken here to use the PCU throughput requirements and not the current PCU capacities, as over-dimensioning of the PCUs will be compounded and result in much more capacity being implemented in the SGSN(s) than is really necessary.

8.4   GPRS Core Network

The dimensioning of the core network must not only take account of the requirements of the radio side, but also of other factors such as the large amount of overhead required by the various protocols described in Chapter 3. We can refer to factors that directly affect the throughput as 'throughput limiters', some of which must be taken into consideration by the planners while others are of interest because they affect the user's perceived data rate – see Section 8.5. The main throughput limiters for both the radio side and core network are listed in Figure 8.7. The diagram also shows the bottleneck caused by the use of 16-kbit TRAU-frames on the Abis interface – this is a throughput limiter in that it prevents the use of CS3 and CS4 on the air interface.

In order to appreciate the importance of the points listed in Figure 8.7, it is necessary to understand the basic principles of planning and dimensioning for data networks, and this is covered in the following sections.

8.4.1   Traffic Types

When considering data traffic we can differentiate between two traffic types: elastic flows, and streaming flows.

Elastic flows are typically from non-real-time applications and can have variable data rates during the transfer. Examples of elastic flows are web-site downloads and file transfers. Such applications typically use TCP for flow control and retransmission, which means there is a feedback loop between sender and receiver. As a result, packet transmission delays result in a lower transmission rate by the sender and unacknowledged packets are automatically retransmitted. This is a so called 'closed loop' system. UDP can also be used within the network if flow control is implemented end-to-end by protocols in the terminal equipment rather than the network.

Streaming flows are typically from real-time applications and require a low packet transmission delay, with as little delay variation as possible (low jitter). Such applications will typically use a real time protocol (RTP) over the UDP transport protocol. UDP is a so called 'unreliable' transport, as no flow control or retransmission is implemented in order to provide faster (low delay) service, i.e. this is an 'open loop' system with no feedback between network elements. Streaming flows require a much higher transmission reliability than elastic flows as there are no retransmissions, e.g. it makes no sense to do retransmissions during a video stream.

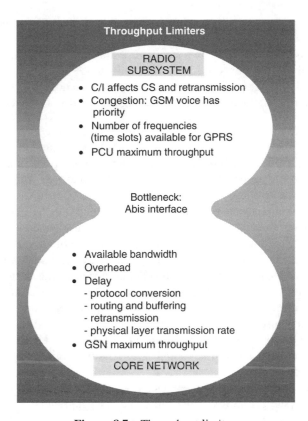

Figure 8.7 Throughput limiters

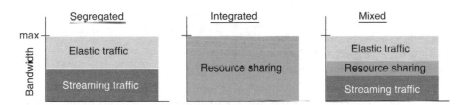

Figure 8.8 Ways of handling different traffic types

When the two data types must be carried by the same network, there are three ways to handle the traffic – it can be segregated, integrated or mixed (as shown in Figure 8.8)

In *segregated* traffic flows, part of the available bandwidth is reserved for streaming flows and the rest is reserved for elastic flows, i.e. there is no sharing, and dimensioning is relatively easy but the resource usage is not efficient.

In *integrated* traffic flows the elastic flow is allowed to utilize fully the bandwidth that is not being used by streaming flows, i.e. high bandwidth utilization is achieved but the dimensioning theory is very complicated and is still evolving.

In *mixed* traffic flows, a certain bandwidth is reserved for elastic flows that is equal to the sum of the guaranteed data rates. The remaining bandwidth can be used by streaming flows, and elastic traffic can make use of any available bandwidth not currently required for streaming flows. To implement such mixed traffic flows a connection admission control (CAC) algorithm is required. CAC is a standard feature on ATM switches, so the mixed traffic solution will be commonly used on ATM links in GPRS (Gn interface) as this will ensure efficient link utilization and guarantee adequate throughput for both data types. It also facilitates the implementation of UMTS using the GPRS core network.

8.4.2 Queueing

Initially, streaming traffic is of little importance as real time applications over the GPRS core network are not foreseen until at least UMTS release 5. This section therefore concentrates on elastic traffic.

The tendency of subscribers to want to make use of the same network resources at the same time means that systems are needed which give all the subscribers access to the required resources while still guaranteeing an acceptable quality of service (QoS), i.e. low transfer time and low blocking probability. Here, the probable waiting time (delay probability) of a packet is an important consideration during the dimensioning process, as the operator must avoid exceeding the maximum delay time required for minimum QoS. This has lead to the development of so called 'queueing models' offering specific parameters that can be used to define a variety of traffic situations. This is necessary as it is possible for elastic traffic to use different kinds of protocols, i.e. UDP is unacknowledged, TCP is acknowledged, and a mixture of both is also possible – and so a number of different models are required.

Queueing models are usually defined using Kendall's notation. Kendall's notation defines a queuing system in terms of the following six parameters:

- the inter-arrival time distribution (A)
- the service time distribution (B)
- the number of servers (c)
- the maximum number of customers (K) $A/B/c/K/m/Z$ = Kendall's notation
- the size of the customer population (m)
- the queue discipline (Z)

In practice, it can very often be safely assumed that there is no limit to the length of the queue, and that there is no limit on the available number of customers, i.e. $K = \infty$ and $m = \infty$. This gives rise to the short-form Kendal notation, e.g. $M/M/1$ – as used later in this text. The other parameters can have the following characteristics:

The arrival process (A) can be:	M = exponential
	D = deterministic
	G = general
The service time (B) can be:	M = exponential
	D = deterministic
	G = general
The servers (c) can be:	1 = one server
	R = R servers

The queue discipline (Z) can be: PS: processor sharing
FIFO: First in first out
FCFS: first come, first served (same as FIFO)
LIFO: Last in first out
LCFS: Last come, first served (same as LIFO)
RSS: Random selection for service
PRI: Priority service

For example, file transfer via FTP uses TCP for secure transmission with flow control, and can be described as $M/G/1$-PS, but WAP uses UDP for shorter transmission times and can be described as $M/M/1$-FIFO. These two examples can not be modelled (or calculated) in the same way as they have different QoS requirements. Depending on which queuing model is used to describe the system, various calculations can be performed to quantify the model.

It should be noted that recently such queuing models have been criticized for not being exact enough. It must be said that such models were developed to provide easily calculable representations of network behaviour, with 'good enough' accuracy. For example the $M/M/1$ model predicts network behaviour with an accuracy of within 10 % of actual observed results, i.e. better than 90 % accuracy. This is deemed by many to be 'good enough' for planning purposes for networks which will in any case later routinely undergo performance analysis and fine-tuning.

8.4.2.1 $M/G/1$-PS Queueing Model for TCP Traffic

In using this model to determine link capacity, time in system, etc., it is assumed that each source can send at the available link rate, which means that the peak throughput available to each subscriber over the air interface can be greater than the available bandwidth on the Gb or IuPS interface (depending on how many other subscribers wish to make use of the link). Hence, processor sharing is necessary and assumes that the available bandwidth will be shared in such a way as to accommodate satisfactorily all simultaneous data transfers. The 'time in system' (sojourn time) for a data packet is equal to the time required for transmission (service time) plus the waiting (queueing) time. The average sojourn time T_x is directly proportional to the service time x but is also dependent on the link utilization ρ. For a service time of x in an $M/G/1$-PS queue, T_x is given by equation (1).

$$T_x = \frac{x}{(1 - \rho)} \tag{1}$$

The time required for transmitting a data packet (service time) x can be calculated as shown in equation (2), where C is the link capacity and l is the length of a file.

$$x = \frac{l}{C} \tag{2}$$

When considering the waiting probability another important factor which must be taken account of is the *utilization* of a link. The utilization factor ρ can be calculated using equation (3). Where λ_e is the mean arrival rate of elastic traffic calls, and x_{mean} is the mean service time.

$$\rho = \lambda_e \times x_{mean} \tag{3}$$

The mean service time can be calculated as shown in equation (4), where l_{mean} is the average file length.

$$x_{mean} = \frac{l_{mean}}{C} \tag{4}$$

The mean response time refers to the time the destination needs to respond to a request from the source, i.e. the 'round trip' time from subscriber to server. The mean response time is independent of the service time distribution and depends solely on the mean value of the service times calculated using equation (3). Hence, by making use of a processor sharing system, any dependence of the average response time on the variance of the service time is removed.

A formula for the desired link capacity can be derived from equations (1), (2) and (3). The formula for the link capacity C_l is shown in equations (5).

$$T_l = \frac{x}{(1-\rho)} \quad \text{where} \quad x = \frac{l}{C_l}$$

Therefore:

$$C_l \times T_l = \frac{l}{(1-\rho)} \quad \text{so} \quad C_l \times T_l - C_l \times T_l \times \rho = l$$

Therefore:

$$C_l \times T_l = C_l \times T_l \times \rho + l \quad \text{where} \quad \rho = \lambda_e \times x_{mean}$$

Therefore:

$$C_l \times T_l = l + C_l \times T_l \times \lambda_e \times x_{mean} \quad \text{where} \quad x_{mean} = \frac{l_{mean}}{C_l}$$

Therefore:

$$C_l \times T_l = l + C_l \times T_l \times \lambda_e \times \frac{l_{mean}}{C_l}$$

Therefore:

$$C_l = \frac{l}{T_l} + \lambda_e \times l_{mean} \tag{5}$$

The capacity C provides for an average sojourn time of T_l in the system for a file of size l. The actual file size l used by the network planners can simply be the anticipated mean file size, or a specific file length (that covers all file sizes up to and including the given length) for which a certain average transmission time has to be guaranteed.

With regard to network dimensioning, each link has to be dimensioned with the required end-to-end network delay. As such, the whole network can be seen as one processor sharing system having a specific sojourn time comprising of a 'service time' and 'waiting time' (delay factor). Again, this delay factor is the result of a further reduction in the service rate due to processor sharing. The actual delay depends on the link capacity C and the number of files n being transferred, i.e. service rate $= C/n$. In effect, each link has the same characteristics as the whole system.

To simplify the modelling, the transmission delay for individual IP packets can be safely ignored. IP packets are generally delivered globally in less than six hops and a 552 byte packet using a link with a capacity of 100 kbit s^{-1} will take 40 ms per hop.

8.4.2.2   *M/G/R*-PS Queueing Model

If there is a limit on the highest rate available for each connection, e.g. due to a limited access rate, then it is more realistic to make use of the *M/G/R*-PS queuing model. Using this model, the number of available channels R depends on the link capacity C divided by the peak rate from the source (the user), i.e. $R = C/r_{peak}$. Using Kendal's notation, R actually stands for the number of servers in the processor sharing queue, but as we are dealing with packet routing, R can be taken here to mean the number of available channels.

Equation (6) can be used to calculate the average sojourn times T_x for a service of length x in respect of the *M/G/R*-PS model. Where E_c represents input from the Erlang C formula, which gives the probability that an arriving packet will have to wait.

$$T_x = \left[1 + \frac{E_c}{R(1 - \rho)} \right] \times x \tag{6}$$

The multiplying factor x is necessary because the sojourn time is proportional to the length of the file being transferred and inversely proportional to the access rate, i.e. as the access rate increases, the sojourn time decreases. As x refers to the time for transferring the file, it is referred to as the 'service' time and can be calculated as shown in equation (7), where r_{peak} is the limited access rate and l is file length (note that x does not include waiting time).

$$x = \frac{l}{r_{peak}} \tag{7}$$

Using the above calculations is a very complex process (especially with regard to the Erlang C formula) so in normal practice a computer based planning tool will be used to simplify the procedure.

As with the *M/G/1*-PS model, each link has to be dimensioned with the required end-to-end network delay and, as such, the whole network can be seen as one processor-sharing system having a specific sojourn time. Again, the additional delay factor ('waiting' time) is the result of a further reduction in the service rate due to processor sharing. The actual delay depends on the link capacity C and the number of files n being transferred, i.e. service rate $= C/n$ where $C/n < r_{peak}$.

As with the *M/G/1*-PS model, each link has the same characteristics as the whole system and the transmission delay for individual IP packets per hop can be ignored.

As both the above models, *M/G/1* and *M/G/R*, use processor sharing, equations (6) and (7) will produce the same results as equations (1) to (4) if there is only a single server (channel), i.e. $R = 1$.

8.4.2.3   *M/M/1* Queueing Model

Assuming that the packet inter arrival times have a Poisson distribution, an intensity of λD arrivals per second and each source transmits single packets with mean size l_{mean}, the dimensioning can be carried out according to the *M/M/1* queueing model. The sojourn time can be calculated as seen in equation (8).

$$T = \frac{x_{mean}}{(1 - \rho)} \tag{8}$$

It is possible to derive the required capacity C from the sojourn time. The formula for the calculation of C is shown in equation (9).

$$C = \frac{l_{\text{mean}}}{T} + \lambda_e \cdot l_{\text{mean}} \qquad (9)$$

As seen in the $M/G/1$-PS model the mean service time can de calculated with the formula in equation (10).

$$x_{\text{mean}} = \frac{l_{\text{mean}}}{C} \qquad (10)$$

With regard to the network dimensioning, the end-to-end network delay is the sum of the total average delay (service time + waiting time) of each interface (including router interfaces). Each link delays the packet and is modelled by a FIFO queue.

8.4.3 Traffic

As mentioned previously, different applications produce different kinds of traffic, which in turn use different protocols. We must therefore consider three different types of traffic:

- acknowledged, reliable traffic: TCP;
- unacknowledged, unreliable traffic: UDP;
- mixed traffic: TCP + UDP on the same link.

8.4.3.1 TCP

The characteristics of TCP traffic are:

- the inter arrival times have Poisson distribution,
- the service time has a general distribution.

Applications such as FTP use TCP/IP. These packets are then compressed, segmented and encapsulated in the mobile for transmission to the SGSN, where the original TCP/IP packets are reassembled. Unfortunately, this IP address refers to the end destination so the core network (Gn interface) is unable to use it for routing the packet. Instead, the GPRS tunnelling protocol (GTP) is used. However, at the time of writing, TCP is not supported on the Gn interface so the TCP/IP packets from the mobile have to be encapsulated and transmitted over UDP as UDP/IP packets. For this reason, later calculations only consider the UDP traffic.

The use of TCP/IP over UDP/IP for file transfers using FTP has the effect of changing the characteristics of the UDP traffic, i.e. instead of a service time with a negative exponential distribution, UDP acquires a general distribution. If FTP traffic becomes a significant percentage of the overall traffic, the queueing model for UDP may have to be reconsidered to account for this.

8.4.3.2 UDP

The characteristics of UDP traffic are:

- the inter arrival times have a Poisson distribution,
- the service time has a negative exponential distribution.

This is because currently the only UDP application is WAP, so the quantity of data transferred is small enough to be carried by a single packet, and the $M/M/1$ queueing model can be used to calculate the bandwidth.

8.4.3.3 Mixed Traffic

This refers to TCP and UDP traffic being transferred via the same link without the use of permanent virtual circuits (PVC) to guarantee bandwidth. As a result, the models shown previously are not always valid for mixed traffic and so a queueing model is difficult to apply. The solution here is to use the weighted average of the TCP and UDP packet sizes to calculate the required capacity.

8.4.4 Routing

An IP network can be thought of in terms of 'links' and 'nodes'. The nodes are the network elements which send and receive packets of data to and from other network elements. The packets travel from one node to another via the 'links'. The nodes read the destination address in the packet header and then decide which node the packet should be sent to next in order eventually to reach the destination. This operation is called 'routing' and the complexities involved in making efficient/effective routing decisions have lead to the development of various protocols and algorithms. The decision about which approach to use must be made during the planning stage as it will be based on:

- quality of service (QoS) requirements,
- capability limitations of the infrastructure,
- throughput capacity of nodes / interfaces,
- network stability.

QoS considerations include end-to-end blocking probability, maximum allowed delay time, and minimum bandwidth requirements. If the network is to take advantage of existing infrastructure, the capabilities of the transmission layer are important in dimensioning the links. The throughput of a node is limited either by its own processing power or by the throughput of the associated interfaces, and is therefore a very important consideration in the dimensioning process. Network stability is improved by a uniform link utilization – a characteristic which is directly influenced by the choice of routing methodology.

The possible routing algorithms are described in the following sections.

8.4.4.1 Static Routing

Static routing is normally only used with simple network topologies. Each router basically has a list (a routing table) which defines a fixed relationship between ports and packet destinations, i.e. all packets for a certain destination go out through a certain port. The devices do not communicate with each other to update each other's routing tables. Instead, the network administrator has to define possible routes and update each router manually. The two major disadvantages of static routing are that it is maintenance intensive in all but trivial networks, and if a node or link is overloaded or fails, routing tables can not be dynamically adapted to make use of alternative routes.

8.4.4.2 Dynamic Routing

The use of dynamic routing algorithms enables routers to maintain up-to-date routing tables automatically, indicating currently available paths through the network. There are a number of dynamic routing protocols available which differ in the way they discover and evaluate the 'new' routes. These protocols can be broadly classified into the following three categories: distance vector protocols, link state protocols, and hybrid protocols.

Distance vector protocols enable each router automatically to maintain an IP routing table based on the distance between itself and every destination on the network. This 'distance' is also referred to as the 'cost' of the path, and enables possible paths to be compared. Routes with lower capacities can be given higher costs to make them less desirable. Costs can also be changed dynamically by the router to account for changes in the network topology. The path with the lowest cost is the one stored in the routing table. In order for 'cost comparisons' to be made, each router regularly 'advertises' its table to its neighbours (typically every 30 seconds). As a result, each router uses the lowest cost path, and changes in the network state, e.g. a node failure, result in all affected routing tables being updated. The period of time required for all routing tables to reflect accurately the current state of the network is called the 'convergence time'. Problems can occur during this period in large complex networks as the network may suffer from inconsistent routing decisions. Generally, the larger and more complex the network, the longer the convergence time, therefore larger networks are more likely to suffer from routing loops and instability during convergence. RIP and BGP are enhanced vector routing protocols designed to reduce the likelihood of network instability during convergence.

The propagation of inaccurate distance vector tables through the network during convergence can result in circular paths known as 'routing loops' being created, i.e. router 'A' believes that router 'B' has a path to the destination, router 'B' believes that router 'C' has a path to the destination, and router 'C' believes that router 'A' has a path to the destination. Routing loops can only exist for a short time before all routing tables are accurately updated, but during this time they contribute to network congestion and overload. A consequence of such loops is the so-called 'count to infinity' problem. In the above scenario, the failed router is no longer sending updates containing the valid route to the destination. The other routers in the loop will therefore receive updates from each other containing a route which is actually no longer valid. The problem arises due to the simple fact that when a router updates its routing table using the distance vectors from neighbouring routers, it must obviously add the cost of the link from itself to the neighbour, i.e. if the neighbour has a link cost of 3 to a certain destination then the router will have a cost of at least 4 for this route. As the routers update each other, they will each in turn add their own link cost to the total and so, with each update, the cost of the route will be increased theoretically to infinity. To combat this problem, limits can be set on the size of link costs, e.g. with RIP link costs can have a maximum value of 16 and links that exceed the maximum are automatically considered invalid.

The so called 'split horizon with poison reverse' is an enhancement that dramatically reduces convergence times. The split horizon approach states that routing information can not leave the router on the same interface through which it was received, i.e. if router 'A' sends information about a certain destination to router 'B', then router 'B' is forbidden to send any information about the same destination to router 'A'. Split horizon was enhanced with the addition of 'poison reverse'. The poison reverse approach states that when a router learns of a route through an interface, it should advertise that route as unreachable back through the same interface. To do this, the router sends an update indicating that the route has an infinite number of hops, e.g. RIP allows a maximum of 15 hops so setting the hop count to 16 would mean 'infinite'. This approach drastically improves convergence times in large, highly redundant, networks as loops and the count to infinity is avoided.

Link state protocols have been developed in response to the need for a routing protocol that deals effectively with the requirements of today's increasingly larger and more complex networks. Link state protocols generate and maintain a database in each router, containing a description of every device in the network and the links between them. This is enabled by the exchange (between routers) of so called link state advertisements (LSA), which are sent out periodically or triggered, for example, by a change in the state of a link (see section on speed of convergence below). An LSA is a list of a router's current links along with the current cost of each link. Each router then uses the SPF (shortest path first) algorithm to calculate the best route to each destination based on the information in the database. SPF produces a tree of 'shortest paths' to all destinations, with its own router as the root.

The routing tables are updated with the results.

Hybrid routing protocols attempt to combine the advantages of both distance vector and link state protocols. Updates are triggered by changes within the network (as with link state protocols) which results in less network overhead than if periodic updates are performed. The metrics used to describe the links are much more detailed than the 'distance' (cost) metric used for distance vectors. Convergence times are usually shorter than for other protocols as updates are event driven and require less overhead.

8.4.4.3   Choice of Routing Protocol

The three main factors affecting the choice of routing protocol are convergence time, stability, and scalability. These factors must be carefully considered by the operator with regard to the new network, as a mistake here can have a serious impact on the performance of the network as a whole.

Speed of convergence is important to all networks, but the larger and more complex a network becomes, the more important it is to have fast convergence as this can directly affect the QoS, i.e. the faster a network converges, the lower the probability of instability. Before a convergence can be initiated, the failure of a node or link must first be detected, and the longer this takes, the more time will be lost before normal operations can be resumed. For example, a node using link state routing must miss several link state advertisements (LSA) before a convergence is initiated. The use of so called 'triggered updates' provides the capability to initiate convergence immediately. When a device detects a failure or changes the cost of a certain route, it immediately updates its neighbours, which are in turn triggered to update their neighbours. Triggered updates are supported by all three types of routing algorithm and result in significantly faster convergence times.

Stability refers to the problems that can occur during convergence, some examples of which are routing loops and the so called 'count to infinity' problem described previously. Routing loops can result when a router becomes unavailable. Distance vector algorithms are more susceptible to such instability. The use of link state or hybrid algorithms reduces the potential for these problems to occur.

Scalability must be considered if significant growth can be expected in terms of size, complexity or both. The scalability of distance vector algorithms is known to be limited in the case of large networks with high redundancy. In such cases, link state or hybrid algorithms should be considered as the use of LSA, link state databases and detailed metrics result in solutions which are readily scalable.

RIP and OSPF are among the most popular routing protocols used by IP devices. The routing information protocol (RIP) is a distance vector algorithm distributed by U.C. Berkeley as part of their 4.x BSD UNIX systems. RIP was adopted by many internet sites without much attention being paid to the weaknesses outlined above. The open shortest path first protocol (OSPF) is a link state algorithm that addresses all the problems associated with RIP. OSPF adopts a logical structure which separates the network into discrete areas so that only routers within an area exchange LSA messages. This reduces both the volume of LSA traffic and the size of the link state database. RIP has enjoyed considerable success but is fast being superseded by OSPF as network operators look for better performance in their networks.

Network operators also make use of static routing in certain circumstances. For example: to define a default route, to define a particular route to a particular host, to provide more efficient resource use on certain routes (no bandwidth used for updates), and to provide a more secure environment (the administrator has control of important links).

8.4.5 Overhead Factor

Overhead is the extra information added to a data packet in the form of headers, that enables various protocols necessary for the operation of the network to be realized. For example, the TCP protocol adds a header to a packet in order to realize a reliable data transfer between nodes; the GTP protocol adds a header to a packet in order to define an individual data flow across an IP network. The disadvantage of adding the overhead is that, in addition to the network capacity required to transfer the user data, the operator must also provide extra capacity to take account of the overhead. Therefore, the amount of overhead must be defined in relation to the amount of user data. To do this, a so called 'overhead factor' must be calculated that can be used as a multiplier to determine the total capacity requirements.

As shown in Chapter 3 it is possible for some interfaces to have various possible implementations for the physical layer (layer 1) and the data link layer (layer 2). This means that the operator must calculate the overhead for the particular implementation chosen in the network being dimensioned. In calculating the overhead, it must also be considered that data packets and certain protocols can have variable lengths, for example, the header for the LLC layer.

The overhead factor for a given interface is calculated by adding the size of the IP packet (P_{SIP}) to the sum of the other overheads, and then dividing the total by the size of the IP packet – as shown in equation (11):

$$\frac{P_{SIP} + \sum_1^i oh_i}{P_{SIP}} \tag{11}$$

The resulting overhead factors can be used as a multiplier during network dimensioning to calculate the actual capacity requirements based on the traffic requirements.

8.4.5.1   Gb Interface Overhead Factor

Table 8.3 shows the protocols present on the Gb interface along with their individual overheads. The 40 bytes for the LLC layer represent the maximum possible header size. It should also be noted that only one encapsulated LLC frame is carried by one FR frame.

Taking a typical IP packet size of 552 bytes and using the formula given above, the overhead factor for the Gb interface can be calculated as follows:

$$\text{Gb interface overhead factor} = \frac{552 + 4 + 40 + 53 + 4 + 8}{552} = 1.1975$$

By using different packet sizes we can create a reference table of overhead factors for the Gb interface. Table 8.4 shows typical packet sizes generated by applications using the TCP and UDP protocols:

The payload size for frame relay is 1600 bytes, which is reflected in Table 8.4 by the low overhead factor associated with the largest packet size of 1500 bytes. Packets larger than this would have to be segmented by the SNDCP layer which would mean that each segment would require the overhead normally used for one packet.

Table 8.3 Overhead from protocols on the Gb interface

Protocol name		Overhead (bytes)
SNDCP	oh_{SNDCP}	4
LLC	oh_{LLC}	40
BSSGP	oh_{BSSGP}	53
NS	oh_{NS}	4
FR	oh_{FR}	8
	TOTAL	109

8.4.5.2   Gn Interface Overhead Factor

In calculating the overhead factor for the Gn interface, allowance must also be made for the so-called 'padding' of ATM cells. ATM cells are 53 bytes long and consist of 48 bytes of payload and 5 bytes of header. If a cell is not completely filled with user data, the unused bytes must be filled with padding bytes. For the calculation, the sum of the overhead is added to the length of the IP packet and the total is divided by 48. If the result is not a whole number of cells then it must be rounded up as it is not possible to send a fraction of an ATM cell.

Table 8.4 Overhead factors from different IP packet sizes on the Gb interface

	Gb Interface					
	TCP (bytes)				UDP (bytes)	
IP packet size	80	552	576	1500	44	300
Overhead factor	2.3625	1.1975	1.1892	1.0727	3.4773	1.3633

There are four possible layer 2 protocols for the Gn interface:

- ATM-AAL5,
- ATM-AAL5-LLC/SNAP,
- Ethernet,
- frame relay.

The overhead for all the protocols found on the Gn interface are given in Table 8.5.

With an IP packet size of 552 bytes and using ATM-AAL5 as layer 2, the required number of ATM cells are given by:

$$\frac{552 + 20 + 8 + 20 + 8}{48} = \frac{608}{48} = 12.67 \text{ rounded up} = 13 \text{ cells}$$

This gives a total, including padding bytes, of:

$$13 \times 48 = 624 \text{ bytes (total padding} = 624 - 608 = 16 \text{ bytes)}$$

The overhead factor for ATM-AAL5 is given by the total bytes divided by the packet size:

$$624/552 = 1.1304$$

Which means that once the capacity required for the initial IP packet has been calculated, it must be multiplied by this factor (1.1304) to account for the overhead.

The overhead factors for the other possible layer 2 protocols can be calculated in the same way and are given in Table 8.6.

Table 8.5 Overhead from possible protocols on the Gn interface

Protocol name	Overhead
GTP	20
UDP	8
IP	20
Ethernet	34
Frame relay	8
LLC/SNAP	8
AAL5	8

Table 8.6 Overhead factors for the Gn interface using different layer 2 protocols

Protocol name	Overhead factor
ATM	1.2482
ATM-LLC/SNAP	1.2482
Ethernet	1.1014
Frame relay	1.8000

It can be seen from the calculations above that the overhead factor for any protocol is inversely proportional to the packet size, i.e. the larger the packet size, the lower the overhead factor.

8.4.5.3   Gi Interface Overhead Factors

The Gi interface has the same layer 2 possibilities as the Gn interface and is therefore calculated in the same way. The calculation is made easier by the fact that Gi is a so called 'pure IP' interface, i.e. GTP and UDP are not necessary so the IP packet passes directly to layer 2.

There is still the problem with padding when ATM is used (see Section 8.4.5.2). The calculation of the overhead factors in Table 8.7 divides the number of bytes to be trans-ferred by the ATM cell size and rounds up to the nearest whole number of cells, i.e. as before, the calculation uses rounding to include the padding.

The overhead factors for the Gi interface are significantly lower than the Gn interface because the tunnelling (GTP) and transport (UDP) protocols are not used on this interface.

Table 8.7   Overhead factors for the Gi interface using different layer 2 protocols

Protocol name	Overhead factor
ATM	1.1522
ATM-LLC/SNAP	1.1522
Ethernet	1.0616
Frame relay	1.0145

8.4.6   Dimensioning the Network

When dimensioning the network, many factors must be taken into consideration. In addi-tion to the required capacity for IP packets and additional protocol overhead, the signalling load, different kinds of traffic and the asymmetric nature of the traffic flow (UL:DL) must also be considered. These factors and others are covered in the following sections.

8.4.6.1   The SGSN

In dimensioning the traffic for an SGSN, the TCP, UDP and so-called 'additional traffic' must be considered separately. The following sections consider the different types of traffic from the Gb, Gn and Iu interfaces which must be handled by the SGSN. The Iu interface to the UMTS radio network must be considered here as UMTS will make use of the GPRS core network. 'Additional traffic' refers to the traffic incurred by signalling, DNS access and SGSN change.

SGSN TCP Traffic
TCP traffic is calculated using the $M/G/R$-PS model and the information (SGSN TCP traffic parameters) required for the calculations is listed below:

- Number of subscribers to be served by the SGSN;
- Number of PVCs for load sharing;

- Spare factor (only required if load sharing is to be implemented);
- Ratio of uplink traffic to downlink traffic;
- Mean traffic per subscriber;
- Percentage TCP traffic in mean traffic (TCP factor);
- Mean file size;
- Peak throughput;
- Overhead factor for the uplink;
- Overhead factor for the downlink.

The above information will have to be calculated, gathered from reliable sources or estimated.

In the simplest scenario, one SGSN will initially serve an entire national network and so the number of subscribers is the percentage of the population that is expected to make use of GPRS. Depending on the handling of the different types of traffic (see 8.4.1), it may be necessary to configure the permanent virtual connections (PVC) for load sharing – in which case a spare factor is defined in case of non-balanced traffic load sharing. The spare factor is typically around 10 %. The asymmetric nature of internet traffic allows us to dimension less capacity in the uplink (UL) than in the downlink (DL). The actual ratio can be based on observed traffic patterns or estimated, e.g. from fixed internet access, as can the values for mean DL traffic per subscriber, the percentage TCP traffic and mean file size. The overhead factors are known from the calculations described in Section 8.4.5. The ratio of UL to DL traffic and the percentage TCP traffic are important as they are used to simplify the capacity calculations, i.e. once the downlink has been calculated we only need to apply the UL:DL ratio to calculate the uplink. The peak throughput represents the maximum possible rate from the subscribers. Some typical values based on internet traffic are 96 kbyte for the mean file size and 56 kbit s^{-1} for the peak throughput. The parameters described above are used in the following to calculate the SGSN TCP traffic.

The estimated downlink TCP traffic per PVC can be found by adding the average downlink TCP traffic per PVC and the spare downlink TCP traffic per PVC, as follows:

$$\text{Est\_DL\_TCP\_per\_PVC} = \text{Ave\_DL\_TCP\_per\_PVC} + \text{Spare\_DL\_TCP\_per\_PVC}$$

As stated above, the spare traffic is calculated as a percentage of the average by using a so-called 'spare factor' (e.g. 0.10), so the downlink TCP traffic per PVC can be estimated by:

$$\text{Est\_DL\_TCP\_per\_PVC} = \text{Ave\_DL\_TCP\_per\_PVC} \times (1 + \text{Spare\_factor})$$

The average downlink TCP traffic will depend on the number of subscribers, the average amount of downlink traffic per subscriber, and the percentage of TCP traffic in the downlink traffic. The average amount of downlink TCP traffic can be calculated as follows:

$$\text{Ave\_DL\_TCP} = \text{Number\_of\_Subs} \times \text{Mean\_DL\_Traffic\_per\_Sub} \times \text{TCP\_factor}$$

Dividing this result by the number of PVCs gives the average downlink TCP traffic per PVC:

$$\text{Ave\_DL\_TCP\_per\_PVC} = \frac{\text{Number\_of\_Subs} \times \text{Mean\_DL\_Traffic\_per\_Sub} \times \text{TCP\_factor}}{\text{Number\_of\_PVCs}}$$

The estimated downlink TCP traffic per PVC (including spare) is therefore given by:

$$\text{Est\_DL\_TCP\_per\_PVC} = \frac{\text{Number\_of\_Subs} \times \text{Mean\_DL\_Traffic\_per\_Sub} \times \text{TCP\_factor}}{\text{Number\_of\_PVCs} \times (1 + \text{Spare\_factor})}$$

Using the above we are able to obtain a value for the estimated downlink TCP traffic per PVC. This value is very important as it enables us to select an initial value for the downlink PVC capacity (Cap_DL_PVC). Ideally, this value would be obtained by calculating back through the Erlang C formula. However, this is impossible so an iterative approach must be used:

- a value for the downlink PVC capacity (Cap_DL_PVC) is assumed;
- calculations for link utilization, delay, waiting probability, etc., are performed;
- the initial value is increased or decreased according to the results;
- the above steps are repeated until an optimum value is found.

The following can be used as a guideline for identifying an optimum value for the downlink PVC capacity:

- the first assumption must always be higher than the estimated downlink TCP traffic per PVC (Est_DL_TCP_per_PVC);
- a recommended starting value is Est_DL_TCP_per_PVC divided by an assumed link utilization (Ass_Link_Util) – a value between 70 % and 80 % should be realistic;
- the optimum value for Cap_DL_PVC should produce an additional transfer time (Add_Trans_T) which is between 4 and 5 % of the minimum transfer time (Min_Trans_T).

To reduce this additional delay, Cap_DL_PVC must be increased as more capacity will decrease the waiting probability, which in turn will reduce the additional transfer time, i.e. if Add_Trans_T is more than 5 % of Min_Trans_T, then Cap_DL_PVC must be increased, BUT if Add_Trans_T is less than 4 % of Min_Trans_T then Cap DL_PVC must be decreased.

Initially then:

$$\text{Cap\_DL\_PVC is set to} = \text{Est\_DL\_TCP\_per\_PVC/Ass\_Link\_Util}$$

The result can then be used to calculate the number of simultaneous subscribers per PVC by dividing the downlink PVC capacity by the peak throughput:

$$\text{Sim\_Subs\_per\_PVC} = \text{Cap\_DL\_PVC/Peak\_Thru}$$

The result will be used later in the Erlang C formula to calculate the waiting probability. The Erlang C formula also requires an estimate for the link utilization based on the downlink PVC capacity:

$$\text{Link\_Util} = \text{Est\_DL\_TCP\_per\_PVC/Cap\_DL\_PVC}$$

The waiting probability can now be calculated using the Erlang C formula:

$$\text{Wait\_Prob}(> 0) = \text{Erlang C}(\text{Sim\_Subs\_per\_PVC}; \text{Link\_Util} \times \text{Sim\_Subs\_per\_PVC})$$

The waiting probability tells us how likely it is that a subscriber encounters a delay when trying to access the network. If this value is very high, it will be necessary to ensure that the duration of the delay is very short. Alternatively, if the probability of a delay is very low then a network operator will accept a higher delay time as very few subscribers will be affected.

The mean delay time can be derived from the minimum transfer time, the waiting probability and the link utilization. First the minimum transfer time needs to be determined as follows:

$$\text{Min\_Trans\_T} = \text{Mean\_File\_Size}/\text{Peak\_Thru}$$

Now, the mean delay time can be calculated:

$$\text{Mean\_delay\_T} = \text{Min\_Trans\_T} \times \text{Delay\_factor}$$

The delay factor is calculated from the waiting probability and link utilization as follows:

$$\text{Delay\_factor} = 1 + (\text{Wait\_Prob}/(1 - \text{Link\_Util}))$$

If the delay factor is low, then the mean delay time will also be low. A low delay factor requires that the waiting probability is low or that the link utilization is high.

We can now calculate a value for the additional transfer time which (when expressed as a percentage of the minimum transfer time) will tell us how Cap_DL_PVC should be adjusted or the next iteration.

$$\text{Add\_Trans\_T} = \text{Mean\_Trans\_T} - \text{Min\_Trans\_T}$$

$$\text{As a percentage} = (\text{Add\_trans\_T}/\text{Min\_Trans\_T}) \times 100$$

When a satisfactory value for Cap_DL_PVC has been found, the calculations for the total TCP capacity for the uplink and downlink can be performed. These values will be used later to calculate the total capacity requirements.

The total uplink TCP traffic is given by:

$$\text{Total\_UL\_TCP\_per\_PVC} = \text{Cap\_DL\_PVC} \times (\text{Traf\_UL}/\text{Traf\_DL}) \times \text{OH\_UL}$$

The total downlink TCP traffic is given by:

$$\text{Total\_DL\_TCP\_per\_PVC} = \text{Cap\_DL\_PVC} \times \text{OH\_DL}$$

These will later be used later in the final dimensioning process.

SGSN UDP Traffic

UDP traffic is calculated using the *M/M/*1 model, i.e. as far as the subscriber is concerned the network appears as a single server without processor sharing.

The following list shows the information (SGSN UDP traffic parameters) required.

- Number of subscribers to be served by the SGSN;
- Number of PVCs for load sharing;
- Spare factor (only required if load sharing is to be implemented);
- Ratio of uplink traffic to downlink traffic;

- Mean downlink traffic per subscriber;
- Percentage UDP traffic in mean traffic;
- Mean file size;
- Average sojourn time;
- Overhead factor for the uplink;
- Overhead factor for the downlink.

Much of the information in this list is the same as for the TCP traffic, i.e. the number of subscribers, load sharing, ratio of UL:DL, the overhead factors and mean traffic per subscriber. Notice however, that 'average sojourn time' now appears in stead of 'peak throughput'. This is because with TCP a processor sharing model was used and so it was necessary to calculate the number of simultaneous subscribers (Sim_Subs_per_PVC) – which depends on the peak throughput (Peak_Thru) – in order to determine the waiting (delay) probability. UDP uses the $M/M/1$ model, which doesn't have processor sharing, so each subscriber can send at link rate and the only delay is given by the sojourn time.

Some typical values for UDP are 300 bytes for the mean file size and 0.2 seconds for the average sojourn time. Obviously the percentage of UDP traffic in the mean traffic will depend on the amount of TCP traffic, i.e. if the mean traffic contains 75 % TCP traffic, then there can only be 25 % UDP traffic.

The parameters described above are used in the following to calculate the SGSN UDP traffic.

The estimated downlink UDP traffic per PVC can be found by adding the average downlink UDP traffic per PVC and the spare downlink UDP traffic per PVC, as follows:

$$Est\_DL\_UDP\_per\_PVC = Ave\_DL\_UDP\_per\_PVC + Spare\_DL\_UDP\_per\_PVC$$

As with TCP, the spare traffic for UDP is also calculated as a percentage of the average using a spare factor, so the downlink UDP traffic per PVC can be estimated by:

$$Est\_DL\_UDP\_per\_PVC = Ave\_DL\_UDP\_per\_PVC \times (1 + Spare\_factor)$$

The average downlink UDP traffic will depend on the number of subscribers, the average amount of downlink traffic per subscriber and the percentage of UDP traffic in the downlink traffic. The average amount of downlink UDP traffic can be calculated as follows:

$$Ave\_DL\_UDP = Number\_of\_Subs \times Mean\_DL\_Traffic\_per\_Sub \times UDP\_factor$$

Dividing this result by the number of PVCs gives the average downlink UDP traffic per PVC:

$$Ave\_DL\_UDP\_per\_PVC = \frac{Number\_of\_Subs \times Mean\_DL\_Traffic\_per\_Sub \times UDP\_factor}{Number\_of\_PVCs}$$

The estimated downlink UDP traffic per PVC (including spare) is therefore given by:

$$Est\_DL\_UDP\_per\_PVC = \frac{Number\_of\_Subs \times Mean\_DL\_Traffic\_per\_Sub \times UDP\_factor}{Number\_of\_PVCs} \times (1 + Spare\_factor)$$

As with TCP traffic, the estimated downlink UDP traffic per PVC (Est_DL_UDP_per_PVC) is used to determine the value of Cap_DL_PVC. With UDP traffic the Erlang C formula is not necessary and the value of Cap_DL_PVC can be calculated directly without the use of estimates and iterations. The value assigned to Cap_DL_PVC depends on the UDP traffic volume. Normally the volume of UDP traffic will be relatively small – amounting to less than 30 % of the total – so Cap_DL_PVC can be set to equal Est_DL_UDP_per_PVC.

The total uplink and downlink UDP traffic can now be calculated as follows: At this point the ratio of uplink traffic to downlink traffic (Traf_UL / Traf_DL), along with the overhead factors (OH_UL and OH_DL) calculated earlier, become important.

The total uplink UDP traffic is given by:

$$\text{Total\_UL\_UDP\_per\_PVC} = \text{Cap\_DL\_PVC} \times (\text{Traf\_UL}/\text{Traf\_DL}) \times \text{OH\_UL}$$

The total downlink UDP traffic is given by:

$$\text{Total\_DL\_UDP\_per\_PVC} = \text{Cap\_DL\_PVC} \times \text{OH\_DL}$$

These will later be used later in the final dimensioning process.

Extra Traffic

This is the so called 'additional traffic' which refers to the extra traffic caused by the following:

- signalling,
- SGSN change,
- DNS access,
- intercepted user traffic.

Of the three interfaces feeding the SGSN, the Gn interface is the most significant in terms of additional traffic. The additional traffic on the Gb interface is contained in the protocol headers and as such has already been accounted for, i.e. no additional calculations are necessary for the Gb interface. The additional traffic on the Iu (UMTS) interface can be calculated with the following simple formula:

$$\text{Traf\_Add\_per\_PVC} = \text{Traf\_Sig\_per\_Sub} \times \text{Subs\_per\_PVC}$$

where:

Traf_Add_per_PVC = additional traffic per PVC(kbit s^{-1})
Traf_Sig_per_Sub = signalling per subscriber(kbit s^{-1})
Subs_per_PVC = number of subscribers per PVC

The important value here is Traf_Sig_per_Sub (signalling per subscriber). This value can be derived from customer behaviour in existing networks but this must be done with caution. The authors have seen values given for this parameter as high as 1 Mbit s^{-1} for every 125 000 subscribers. There is a great danger of over-dimensioning the network when making such assumptions so care must be taken to ensure the values are realistic.

The additional traffic on the Gn interface is calculated according to the following:

Assuming that load sharing will be used, the number of subscribers per PVC has to be calculated with reference to the spare factor for load sharing mentioned previously in the parameters for TCP and UDP traffic. Therefore:

$$\text{Subs\_per\_PVC} = (\text{Number\_of\_Subs}/\text{Number\_of\_PVCs}) \times (1 + \text{Spare\_factor})$$

The signalling traffic per subscriber (Traf_Sig_per_Sub) includes PDP context activation, PDP context deactivation, and routing area changes, and will typically use a packet size of 512 bytes. The signalling traffic per subscriber in the busy hour is given by the equation:

$$\text{Traf\_Sig\_per\_Sub} = \frac{\text{Act} + \text{Deact} + \text{RAC}}{3600} \times \text{Size} \times \frac{8}{1000} \quad (\text{kbit s}^{-1})$$

where:

Act   = number of PDP context activations per subscriber per busy hour
Deact = number of PDP context deactivations per subscriber per busy hour
RAC   = number of routing area changes per subscriber per busy hour
Size = IP packet size

At the time of writing, the authors saw calculations based on the following values. Again it should be noted that such values will depend on subscriber behaviour and will not be the same for all networks. They are provided here only as an indication as to what the reader might expect to find:

Act   = 0.25 PDP context activations per subscriber per busy hour
Deact = 0.25 PDP context deactivations per subscriber per busy hour
RAC   = 1 routing area change per subscriber per busy hour
Size = 512 bytes

Hence, the busy hour signalling traffic per PVC is given by:

$$\text{Traf\_Sig\_per\_PVC} = \text{Traf\_Sig\_per\_Sub} \times \text{Subs\_per\_PVC}$$

Additional traffic is also incurred on the Gn interface when a subscriber changes SGSN, as data is sent from the 'old' SGSN to the 'new' SGSN during the change over. The amount of traffic (Traf_ch_SGSN) depends on number of SGSN changes per subscriber per busy hour (SGSN_ch_per_Sub), the peak throughput (Peak_Thru), the change over time (Change_T) and the number of simultaneous subscribers per PVC (Sim_Subs_per_PVC), and is given by the equation:

$$\text{Traf\_ch\_SGSN} = (\text{SGSN\_ch\_per\_Sub}/3600) \times \text{Peak\_Thru} \times \text{Change\_T}$$

$$\times \text{Sim\_Subs\_per\_PVC}$$

Traffic to and from the domain name server (DNS) is also considered as additional traffic. The amount of DNS traffic (Traf_DNS) is simply the number of requests and responses made per subscriber during the busy hour. This given by the equation:

$$\text{Traf\_DNS} = (\text{DNS\_per\_Sub\_per\_Hr}/3600) \times \text{Size} \times 8 \times \text{Subs per PVC} (\text{kbit s}^{-1})$$

where:

DNS_per_Sub_per_Hr = DNS requests and responses per subscriber per busy hour
Size   = IP packet size
Subs_per_PVC   = number of subscribers per PVC

Again, a value for DNS requests/responses per subscriber per busy hour of 0.1 can be given by the authors as an indication.

Intercepted traffic (Traf_Int) is sent to the relevant law enforcement agency (LEA) from the SGSN via the Gn interface and a border gateway (BG). Both the uplink and downlink are intercepted, copied and sent to the LEA, hence, uplink and downlink traffic for UDP and TCP must be considered. The amount of traffic depends directly on the number of intercepted subscribers. For dimensioning purposes, unless there are special circumstances, the percentage of intercepted subscribers will be very small. Hence an intercept factor (Int_factor) of around 0.03 (3 %) can be given. The additional traffic is given by the equation:

$$\text{Traf\_Int} = (\text{Traf\_UL} + \text{Traf\_DL}) \times \text{Int\_factor}(\text{kbit s}^{-1})$$

where:

Int_factor = represents the percentage of intercepted subscribers
Traf_UL   = uplink traffic (Tot_UL_TCP_per_PVC + Tot_UL_UDP_per_PVC)
Traf_DL   = downlink traffic (Tot_DL_TCP_per_PVC + Tot_DL_UDP_per_PVC)

Total SGSN Traffic

All the values needed to calculate the total required SGSN capacity have been determined in the previous sections. These values must be added together and multiplied with the desired spare factor as follows:

The first step is to calculate the sum of the uplink and down link traffic, for TCP and UDP (including the spare capacity) per PVC.

$$\text{Tot\_Traf\_UL} = (\text{Tot\_UL\_TCP\_per\_PVC} + \text{Tot\_UL\_UDP\_per\_PVC}) \times (1 + \text{Spare\_factor})$$

$$\text{Tot\_Traf\_DL} = (\text{Tot\_DL\_TCP\_per\_PVC} + \text{Tot\_DL\_UDP\_per\_PVC}) \times (1 + \text{Spare\_factor})$$

The next step is to calculate the sum of the additional traffic for the uplink and the downlink (including the spare capacity) per PVC.

$$\text{Tot\_Traf\_UL\_Add} = (\text{Traf\_Sig\_per\_PVC} + \text{Traf\_ch\_SGSN} + \text{Traf\_DNS} + \text{Traf\_Int})$$

$$\times (1 + \text{Spare\_factor})$$

$$\text{Tot\_Traf\_DL\_Add} = (\text{Traf\_Sig\_per\_PVC} + \text{Traf\_ch\_SGSN} + \text{Traf\_DNS})$$

$$\times (1 + \text{Spare\_factor})$$

Finally the total uplink and downlink capacity requirements per PVC can be calculated by summing the above results.

$$\text{Tot\_Capacity\_UL\_per\_PVC} = \text{Tot\_Traf\_UL} + \text{Tot\_Traf\_UL\_Add}$$

$$\text{Tot\_Capacity\_DL\_per\_PVC} = \text{Tot\_Traf\_DL} + \text{Tot\_Traf\_DL\_Add}$$

The values calculated using the above can then be used to determine the actual equipment requirements.

8.4.6.2 The GGSN

The dimensioning of the GGSN is very similar to the dimensioning of the SGSN. The key differences being:

- the overhead for the Gi interface must be calculated;
- only the Gn interface has additional traffic;
- interception is only possible at the SGSN so no additional traffic for GGSN;
- PDP activation is driven by SGSN so no additional traffic for GGSN.

As with the SGSN, TCP and UDP traffic (including the protocol overhead and spare factors) have to be calculated separately and summed to determine the total capacity required for the Gn and Gi interfaces. The process uses the same formulas and parameters used in the SGSN calculations above.

8.4.6.3 The Border Gateway

When dimensioning the border gateway (BG) we are concerned with the capacity of the Gp interface. The BG capacity is calculated in the same way as the SGSN but also takes into consideration the protocol overhead incurred by the Gp interface protocol stack. There is no additional traffic unless:

- roaming traffic is crossing the border gateway;
- subscribers are using a service from outside the gateway (thus producing DNS traffic);
- the gateway is used for interception.

As with the SGSN, TCP and UDP traffic (including the protocol overhead and spare factors) have to be calculated separately and summed to determine the total capacity required for the Gp interface. The process uses the same formulas and parameters used in the SGSN calculations above.

8.4.6.4 Cross-checking

A cross-check looks at the traffic between SGSN, GGSN and border gateway relative to the uplink and downlink of the SGSN to see if the capacities are balanced.

The equations for the cross check are as follows:

$$\text{downlink:} \quad \sum \text{SGSN} + \sum \text{GGSN-to-BG must equal} \sum \text{GGSN} + \sum \text{BG-to-SGSN}$$
$$\text{uplink:} \quad \sum \text{SGSN} + \sum \text{BG-to-GGSN must equal} \sum \text{GGSN} + \sum \text{SGSN-to-BG}$$

The cross-check presupposes that:

- the SGSN downlink traffic comes partly from the GGSN and partly from the BG;
- the SGSN uplink goes partly to the GGSN and partly to the BG.

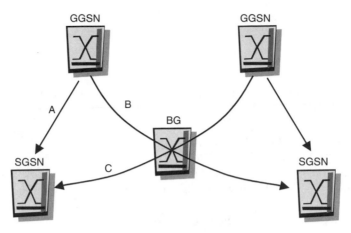

Figure 8.9   Downlink cross-check

The downlink cross-check scenario is illustrated in Figure 8.9.

Hence:

$$\text{downlink}: \sum \text{SGSN} + \sum \text{GGSN-to-BG} = \sum \text{GGSN} + \sum \text{BG-to-SGSN}$$

$$(A + C) + B = (A + B) + C$$

A slight imbalance in the cross check can be ignored but if the results are seriously out of balance, the previous calculations must be checked and corrected.

8.4.6.5   Phases

When a network operator plans a new network it must consider the future growth of that network. With time, the number of subscribers will grow as will the number and diversity of services offered, i.e. more users will have more ways to use the network. Obviously then, the capacity requirements in the short, middle and long term will increase therefore additional infrastructure will be required.

For the above reason, the planning and dimensioning of new networks usually adopts a phased approach. For example, in addition to the initial network requirements, the network operator will also give the equipment supplier or consultant information on the anticipated future requirements. The supplier will then plan and dimension the network for the initial requirements which will be Phase 1 of the implementation. The supplier will then perform further planning for Phase 2 and Phase 3 which might start 18 months and 36 months after Phase 1, respectively. This forward planning of Phases 2 and 3 will only be as good as the predictions provided by the network operator but it serves several useful purposes. It allows the operator to:

- set milestones for network growth;
- to check predictions (models and simulations) with real-world development;
- to make cost comparisons between different suppliers;
- to adjust the timing of new development to suit the real demand.

While Phase 1 will depend to a certain extent on assumptions, later phases also depend on assumptions for predicted growth and utilization. With this in mind the importance of the last point in the above list becomes apparent. If the operator has overestimated the growth then the start of Phase 2 can be postponed. However, the consequences are very different if the operator has underestimated the growth. In this case, the existing infrastructure will have to cope with higher loads with higher offered traffic, which will result in higher blocking. It may be possible to initiate Phase 2 earlier so as to have more capacity available before the situation becomes critical but this is not an ideal solution as the dimensioning for Phase 2 will also probably have to be increased to deal with the higher-than-anticipated demand.

It is clear then that both the approach to dimensioning and the data on which it is based need to be as reliable as possible to avoid such scenarios. It is tempting to deliberately over-dimension a network as this can be justified by saying 'sooner or later the capacity will be used', and while this may be true it has the negative effect of increasing the initial investment required to implement the network. Thus the suppliers who provide cost estimates based on the dimensioning process face a dilemma: if they over-dimension the network they will guarantee QoS but their offer will be expensive, or if they try to be very exact to keep the costs down they risk QoS problems if demand is higher than expected.

8.5 User Aspects

Factors that limit the data rate perceived by the user are those which affect the rate at which data is delivered to the application layer, e.g. to a web browser. We have to differentiate between 'actual transfer rate' and 'perceived transfer rate' as the PCU effectively hides the network delays from the air interface, i.e. the PCU performs statistical multiplexing of users onto the Um and only allocates resources if there is something to send. Perceived data rate limiters are shown in Figure 8.10.

The impact of time slot sharing and cell reselection depend on individual user behaviour, e.g. performing large downloads while mobile. The volume of retransmission depends mainly on C/I and to some extent on mobility.

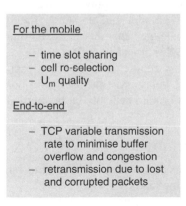

Figure 8.10 Perceived data rate limiters

Due to the 'bursty' nature of data traffic, time slot sharing should only cause a problem when there are not enough resources available. This problem can largely be ignored as initially there will be so little GPRS traffic volume and, as the networks mature, flexible time slot allocation will help to ensure adequate availability of resources. Further, the introduction of QoS (quality of service) support will ensure that those subscribers who need fast service (priority) will be able to get it. The introduction of CS3 and CS4 will also alleviate congestion by reducing data transfer times when the quality of the air interface is good.

Cell reselection (changing the serving BTS – similar to a handover in GSM) is only really a problem for time critical data transfers. The problem here is that during the cell reselection procedure, the mobile cannot send or receive data as it is busy exchanging signalling with the PCU via the BTS. In some situations, the mobile will select the new cell, in others the network will tell the mobile which cell to use. In either case, data transfer will cease and only recommence after the reselection has been completed. The mobile must then request retransmission of the packets it missed. For a file transfer this would not matter but for any time sensitive application, this pause will be a problem.

Retransmission is also necessary when, due to interference on the air interface (Um), some bits in the radio burst have been corrupted (more than can be corrected with FEC) and the entire radio block must be sent again. As the amount of retransmissions increase the PCU will instruct the BTS and MS to use more robust coding schemes which will result in a lower data rate for the user.

On the network side, one of the methods used to reduce core network congestion, particularly buffer overflow, is variable rate packet transmission. The TCP layer will send the data received from the application with an ever increasing data rate until it receives a buffer overflow warning from the network. It will then start decreasing the rate until retransmissions stop, whereupon it will start increasing the rate again. In this way the average throughput is higher than if fixed rates are used. Unfortunately, the user perceives that the rate rises to a maximum and then decreases – leading to the assumption that performance is poor when actually the performance is being optimized in real time.

Although many protocol diagrams for GPRS show that it is also possible to use TCP within the core network (Gn interface), in reality only UDP is usually used. UDP has no flow control and so packet loss/corruption occurring between SGSN and GGSN results in an end-to-end retransmission as TCP can only be utilized between the application layers.

For transmission between the mobile and the SGSN, the SNDCP layers store all packets until they receive an acknowledgement of successful transmission, i.e. retransmission occurs between MS and SGSN, not end-to-end.

It should be clear from this section that adding capacity to the radio side will not necessarily result in reduced delays for the user, and conversely, that low data rates experienced by the user can not automatically be blamed on poor network dimensioning.

8.6   Indoor Radio Networks

In recent years, the provision of radio coverage within buildings has become not only expected, but in some cases mandatory.

Many approaches have been developed to overcome the extremely poor propagation characteristics that exist within enclosed spaces. If you imagine what happens to the mirror flat surface of the water a swimming pool when someone jumps in, the result is identical to what happens to radio transmissions within a building. The radio waves propagate in a circular front away from the source, they are reflected from the flat surfaces (especially those containing metal) and then, because they have the same frequency and similar amplitude, they interfere with each other positively and negatively, which attenuates the signal. Add to this the high density of the building materials and it becomes clear that signals travelling from a mobile within a building to a BTS outside, will suffer from very high losses.

Possible solutions include using small indoor BTS mounted on ceilings and walls, and the use of both passive and active repeaters. Such solutions can utilize networks of individual antennas or lengths of so called 'leaky feeder' (or 'leaky coax'). The former is better suited to rooms and open spaces while the latter is best used for corridors.

In order to justify the expense involved in implementing an internal antenna network there must be a high volume of traffic, i.e. the building represents a so called 'hot spot' within a normal macrocell. Such buildings include:

- shopping centres,
- office buildings,
- exhibition centres,
- airports,
- main rail and underground stations.

All the above are high-traffic areas and the nature of the buildings means that the coverage from a normal macrocell will be poor owing to the path losses. Thus an indoor network becomes an attractive solution.

From the point of view of GPRS, buildings in which a lot of people do a lot of waiting, e.g. airports and stations, become potential future hot spots as the waiting time will be spent 'on-line' rather than reading books or drinking coffee. As such, the use of indoor antenna networks will need to increase as demand for mobile services expands.

The following figures illustrate various means of implementing indoor network coverage the advantages of each and implications for radio planning will be mentioned.

The passive repeater illustrated in Figure 8.11 represents the simplest approach to increasing indoor radio coverage. This technique relies on the building having a nearby source and short indoor cable lengths. Proximity to the BTS is necessary as the uplink cannot tolerate as much loss as the downlink. Hence, the second point, short cable lengths, is important as signal loss increases with cable length (a typical value being 4 dB loss per 50 m). This solution is applicable where an indoor area is completely screened from the external source. Such solutions tend to be limited to a single indoor antenna.

The active repeater with splitter illustrated in Figure 8.12 can be used when the building is too far from the nearest BTS to use a passive repeater. Here the antennas are attached to an amplifier which boosts the signals in both the uplink and downlink. This enables multiple indoor antennas to be connected to a single outdoor antenna. Cable losses are not so important as they can be compensated by the booster. This makes the active repeater a much more flexible solution. However, because the signal from the BTS is being boosted and re-broadcast, care must be taken in the planning stage with regard to interference. This is especially important in buildings such as hospitals where interference with sensitive

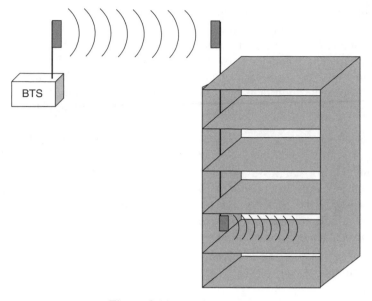

Figure 8.11   Passive repeater

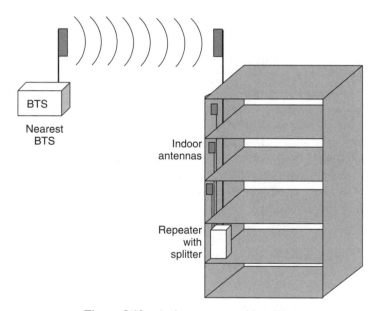

Figure 8.12   Active repeater with splitter

medical instruments can be dangerous to patients. Further, if the signal delivered to the
indoor antennas is too strong, the radio waves will not be sufficiently attenuated within
the broadcast area and antennas in different rooms and floors will interfere with each
other. This means that the repeater must be set up such that the power control performed
by the BTS results in the minimum possible transmission levels.

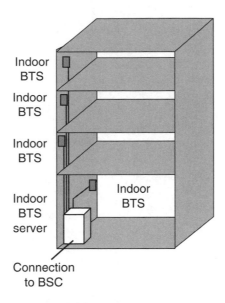

Figure 8.13   Indoor BTS network

The indoor BTS network, illustrated in Figure 8.13, can be used in any building in which traffic volume justifies the expense. As each of the indoor BTS constitutes a cell in its own right, normal cell planning procedures with regard to frequency reuse distances and minimum interference levels must be applied. This solution requires that all the indoor BTS are connected to a server which is in turn connected to a BSC.

One of the main advantages of the indoor BTS network is that several BTS can be placed in the same room, which increases the available capacity for that room only without significantly increasing the interference in the surrounding area. For buildings served by a repeater, the only way to increase the available capacity for a certain room is to increase the capacity of the whole cell, i.e. add TRXs to the nearest BTS. This is obviously not ideal as the increase is not specific for a certain area and it also increases the possibility of interference in neighbouring cells.

From the point of view of GPRS, for which demand is expected to increase steadily, the use of an indoor BTS network is preferable both for the reasons stated above and because such low power BTS are the best solution for covering indoor hot spots without adding complexity to interference-limited networks in cities.

9

Towards All-IP Networks

9.1   The TCP/IP Protocol Suite

9.1.1   The Role of IP

The Internet protocol (IP) is the basic protocol in the TCP/IP architecture. All computers in the Internet world are able to understand IP. It is a connectionless and unreliable protocol, which doesn't use acknowledgements or error correction procedures. The result is a high data rate, but reliability and error correction must be implemented in the upper (higher) protocol layers. The most important tasks of IP are addressing of hosts via the IP addresses (see Sections 9.1.3 and 9.1.5) and the fragmentation/reassembly of packets. The use of a logical architecture with IP addresses allows a flexible routing procedure. The route between two communication partners is not a fixed route from end-to-end, but is a flexible process that always looks for the best route between the two partners. This flexible process can be described as a hop-by-hop mechanism between different stations (routers) which are on the route between the two communication end points. An IP datagram consists of two parts, the header part (see Sections 9.1.2 and 9.1.4) and the data part (which is often called the 'payload'). The header includes various fields, such as where the datagram should be forwarded to, how large the datagram is, and other information that is necessary to offer a 'best effort' service for the packet transmission.

9.1.2   The IPv4 Header

The header of an IP datagram consists of at least five words, and each word is represented by one row containing 32 bits. The fields in the header (illustrated in Figure 9.1) have the following functions.

- *Version*: This field has a length of 4 bits and specifies the IP version of the following packet. With the current IP version 4, this field contains the binary value 0100.
- *Internet header length (IHL)*: This field has a length of 4 bits and defines the length of the header as multiples of 4-octet (i.e. 32-bit) words. Due to the minimum header length of 20 octets (five 32 bit words) this field has the minimum binary value 0101 (decimal 5 = binary 0101).

GPRS Networks.   G. Sanders, L. Thorens, M. Reisky, O. Rulik, S. Deylitz
© 2003 John Wiley & Sons, Ltd.   ISBN: 0-470-85317-4

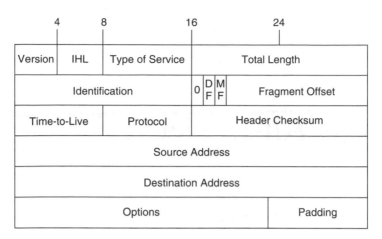

Figure 9.1   The contents of an IPv4 header

- *Type of service (TOS)*: This field has a length of 8 bits and enables the originating host to request different classes of services for transmission of packets. In this field, a service priority (0–7) can be specified, or route optimizations related to costs, delay, throughput or reliability can be requested.
- *Total length*: This field has a length of 16 bits and indicates the length (in bytes or octets) of the entire packet, including header and data. Using all 16 bits, the maximum size of an IP packet is therefore 64 kb (65535 bytes). However, in practice, the packet size is limited by the largest size a specific transport medium can carry, i.e. the maximum transfer unit (MTU). For example, on a token ring network the largest packet size is 4202 bytes, whereas on an Ethernet (802.3) network the maximum packet size is 1518 bytes.
- *Identification*: This field has a length of 16 bits and is used when a packet needs to be fragmented into smaller parts because it is too large to be transmitted whole. This identifier is assigned by the transmitting host to aid in subsequent reassembly of a fragmented datagram. Each fragment of a datagram has the same identification number.
- *Flags*: This field has a length of 3 bits and contains three control flags. The first bit is not used and is therefore always set to 0. The second bit is the 'don't fragment' (DF) bit. A zero means that fragmentation is allowed but if the bit is set to 1, then fragmentation is not allowed. The third bit is the 'more fragments' (MF) bit. If this MF bit is set to zero, then this is the last fragment of a datagram, and a value 1 means that additional fragments related to this datagram will arrive.
- *Fragment offset*: This field has a length of 13 bits and shows the position of the fragment in the original packet. If this is the first packet of a fragment stream or the only packet of a datagram then the offset is zero. In subsequent fragments, this field indicates the offset in increments of 8 bytes.
- *Time-to-live*: This field has a length of 8 bits and indicates the time (in seconds) for which the datagram is allowed to remain in the network before it is discarded. In theory, each router processing this datagram is supposed to subtract its processing time from this field but in practice, a router processes the datagram in less than 1 second. As a result, the router always subtracts one from the value in this field. Therefore the TTL

becomes a hop-count metric rather than a time metric. When the value reaches zero, it is assumed that this datagram has been travelling in a closed loop and is discarded.

- *Protocol*: This field has a length of 8 bits and specifies the higher layer protocol used for transferring this packet. Examples are TCP (6), UDP (17) or ICMP(1).
- *Header checksum*: This field has a length of 16 bits and contains information to make sure that the IP header is error free. The checksum must be recalculated by each router as the TTL field in the header changes after every hop.
- *Source address*: This field has a length of 32 bits and indicates the IP address of the host sending the packet.
- *Destination address*: This field has a length of 32 bits and indicates the IP address of the host which is the intended receiver of the packet.
- *Options*: This field has a variable length and is used for additional options, such as specified source routing or security indication.

9.1.3 The IPv4 Addressing Concept

The IPv4 address format is based on 32 bits, divided into four 8-bit integers, separated by a dot (Figure 9.2). Each of these 8-bit integers is represented by a decimal number between 0 and 255. Hence, this is referred to as 'dotted decimal notation'.

The following is a typical example of an IPv4 address:

IPv4 Address: 212.227.119.105

An IPv4 address can be divided up into two logical parts: the network address and the host address.

The *network address* is related to a network segment and is the part of an IP address that defines the destination network. The *host address* defines a particular host inside

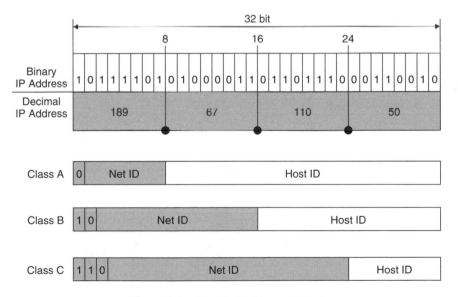

Figure 9.2 The IPv4 address concept

a specific network. The IPv4 addresses are grouped into three main classes, depending on the number of bits used for the network address and host address. (There are two additional classes, but they are not used in commercial networks.)

Class A: An IPv4 address of Class A always starts with a '0' in the first octet of the address. Only the first octet is used as network address, the other three octets are used for the host address. Therefore 128 (starting from 0 up to 127) Class A networks are available, each with a maximum number of $16\,777\,216$ (2^{24}) unique host identifiers.

Class B: An IPv4 address of Class B always starts with a bit sequence of '10' in the first octet of the address. In this case the first and the second octet are used for the network address, and the other two octets are used for the host address. The address range for Class B networks is from 128.0.0.0 to 191.255.0.0. Therefore $16\,384$ Class B networks are available, each with a maximum of $65\,536$ (2^{16}) unique host identifiers.

Class C: An IPv4 address of Class C always starts with a bit sequence of '110' in the first octet of the address. In this case the first three octets are used as the network address, and the last octet is used for the host address. The address range for Class C networks is from 192.0.0.0 to 223.255.255.0. Therefore $2\,097\,152$ Class C networks are available, each with a maximum of 256 (2^8) unique host identifiers.

Table 9.1 shows a comparison of the different classes of IPv4 addresses. An important part of the class concept in IPv4 is the possibility of performing 'sub-netting'. The creation of sub-networks and the use of so called 'sub-net masks' allows large networks to be split up into smaller parts. The sub-netting concept is based on the division of the host ID part of an IP address into two parts, the sub-net number and a host number. The combination of sub-net number and host number is defined as 'local address' or 'local part'. The combination of net ID and sub-net number defines a sub-net address (Figure 9.3).

Table 9.1 A comparison of the classes of IPv4 address

CLASS	Leading bits in the first octet	Net ID bits	Number of possible nets	Host ID bits	Number of possible hosts	Range of values
Class A	0	7	126	24	$16\,777\,214$	1 to 126 (127 is used for internal loopback testing
Class B	10	14	$16\,382$	16	$65\,534$	128 to 191
Class C	110	21	$2\,097\,150$	8	254	192 to 223

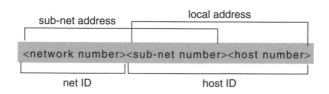

Figure 9.3 The parts of an IP address in case of sub-netting

IP Address:	091.144.107.015	01011011. 10010000. 01101011. 00001111
Sub-net mask:	255.192.000.000	11111111. 11000000. 00000000. 00000000
Network part of the address:	091.128.000.000	01011011. 10000000. 00000000. 00000000
Computer part of the address:	000.016.107.015	00000000. 00010000. 01101011. 00001111

Figure 9.4   The concept of sub-net masks

Since a router must distinguish between (sub-)network part and host part, the standard classification of Class A, B, C network addresses can not be used to differentiate between individual sub-nets. Therefore sub-net masks are introduced (Figure 9.4). Sub-net masks allow certain bits in the host part of an IP address to be used as part of the sub-net address, and can be used to filter out the host number. As with the IP address, the sub-net mask consists of 4 bytes, separated by dots in its written representation. The principle of operation of a sub-net mask can be explained by considering the binary format of the mask. Each bit that is set to '1' in the sub-net mask belongs to the network part of the IP address; the other bits of the IP address, which are used for the host part, are set to '0' in the mask.

9.1.4   Migration from IPv4 to IPv6

The current IP version 4 (IPv4) and the related addressing concept was designed to work with 4 billion IP addresses. For the near future, when more and more electronic devices will require their own IP addresses, a new address concept, offering a much larger number of IP addresses, is necessary. IP version 6 (IPv6), also known as IPng (next generation), is a new protocol based on the advantages of IPv4 and the experience gained from existing IP networks. IPv6 increases the number of available IP addresses enormously and offers a backward compatibility to IPv4, which guarantees a parallel existence of IPv4 and IPv6 during the transition period.

In IPv6, some of the header fields from IPv4 are no longer used. The header checksum is no longer needed, because this task can be done by lower layer protocols. The hop-by-hop segmentation is also removed and replaced by a procedure called path MTU (maximum transfer unit) discovery. In this case, all the hosts will learn the maximum acceptable segment size of the network path. If a host wants to send a larger packet, the network will reject the packet. Therefore the segmentation control flags and the fragment offset field are also no longer necessary.

Some new fields introduced in IPv6 are used to optimize real-time traffic. Such fields will be necessary in converged networks – especially in telecommunications – and are described in the following section.

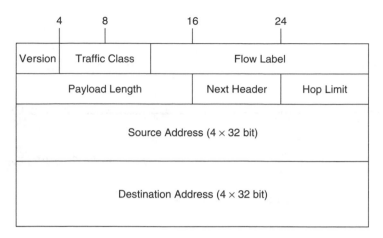

Figure 9.5 The contents of an IPv6 header

9.1.5 The IPv6 Header

The header of an IPv6 packet has a total length of 40 bytes which can be viewed as a general 64-bit header and two 128-bit fields for the source and destination addresses. The complete header is illustrated in Figure 9.5 and consists of the following fields:

- *Version*: This field has a length of 4 bits and specifies the IP version of the following packet. With the IP version 6, this field contains the binary value 0110.
- *Traffic class*: This field has a length of 8 bits and allows applications to specify a certain priority for the traffic they generate, thus introducing the concept of class of service. This enables the prioritization of packets, as in differentiated services. The class field was designed to support real-time traffic.
- *Flow label*: This field has a length of 20 bits and is used to enable faster routing decisions to be made by recognizing packets that need to be handled in the same way. The handling requirement for a particular flow label is known as the 'state' information; this is cached at the router. When packets with a known flow label arrive at the router, the router can efficiently decide how to route and forward the packets without having to examine the rest of the header for each packet. The flow label field was designed to support real-time traffic.
- *Payload length*: This field has a length of 16 bits and defines the length of the packet in bytes (excluding this header). If length is greater than 64 kb, this field is 0 and an optional header (jumbo payload) gives the true length.
- *Next header*: This field has a length of 8 bits and indicates the type of header immediately following the basic IP header. It may indicate an IP optional header or an upper layer protocol. The next header field is also used to indicate the presence of extension headers, which provide the mechanism for appending optional information to the IPv6 packet.
- *Hop limit*: This field has a length of 8 bits and is similar to the IPv4 TTL field, but the value is now measured in hops and not in seconds. The packet is discarded once the hop limit has been decremented to 0.

- *Source address*: This field has a length of 128 bits and indicates the IPv6 address of the host that sent the packet.
- *Destination address*: This field has a length of 128 bits and indicates the IP address of the host that is the intended receiver of the packet.

9.1.6   The IPv6 Address Concept

The IPv6 address concept enlarges the address format from 32 bits to 128 bits. Therefore many more hosts can be defined and more levels of hierarchy are possible (IPv4 only allows the hierarchy elements: net, sub-net, and host). The IPv6 address notation is based on 128 bits, divided into eight 16-bit integers, separated by colons. Each of these 16-bit integers is represented by four hexadecimal digits and is illustrated in Figure 9.6.

The following is a typical example of an IPv6 address:

IPv6 Address: ABCD:EF54:9854:1023:FDEC:AB31:5363:7733

There are some rules that make the notation of IPv6 addresses a little easier.

- All leading zeros in each hexadecimal component can be skipped, e.g. component '0000' can be reduced to '0' or '00A3' to 'A3'. For example:

2490:0000:0000:0000:00B1:0700:200C:4171

can be written as

2490:0:0:0:B1:700:200C:4171

- Inside an address two colons can replace a series of consecutive null 16-bit numbers. This is known as the 'double-colon' convention. This convention can only be used once inside an address.
Example 1:

2490:0000:0000:0000:00B1:0700:200C:4171

can be written as

2490::B1:700:200C:4171

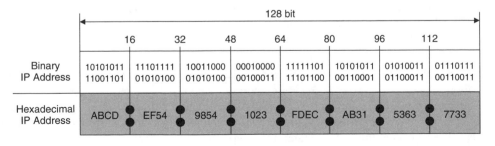

Figure 9.6   The IPv6 address concept

Example 2:

$$0000:0000:0000:ABCD:1234:0000:0000:0000$$

can be written as

$$::ABCD:1234:0:0:0 \qquad or \qquad 0:0:0:ABCD:1234::$$

The backward-compatibility between IPv6 and IPv4 is realized with the conversion of the 32-bit IPv4 address into the new IPv6 address by adding 96 zeros to the beginning of the address. The following is an example of an IPv4 address expressed in the form of IPv6:

Notation of an IPv4 address in IPv6 format   178.124.56.10 can be written as:

$$::178.124.56.10 \qquad or \qquad ::B27C:380A$$

9.1.7 Transmission Control Protocol (TCP) and User Datagram Protocol (UDP)

There are two host-to-host layer protocols that are working on top of the Internet protocol. These protocols are the transmission control protocol (TCP) and the user datagram protocol (UDP). The following sections give an overview of the functionalities of these two protocols along with the similarities and differences between them.

9.1.7.1 Transmission Control Protocol

TCP is a connection-oriented communication service that is responsible for the establishment and termination of virtual circuits. Here 'connection-oriented' indicates only that the service has fixed start and end points – it does not mean that an end-to-end connection is set up like GSM. Additional functions of TCP are:

- *sequencing*: packets are buffered and sorted into the correct order at the destination node;
- *flow control*: packets buffered in the sending node are erased when the receiver sends an acknowledgement (the Ack also indicates how many bytes can be received without overload);
- *error correction*: retransmission of missing packets is requested automatically.

The data unit in TCP is a 'segment' and the header attached to such a segment (illustrated in Figure 9.7) consists of the following fields:

- *Source and destination port*: These fields, each having a length of 16 bits, define the source and destination ports for an end-to-end connection and the higher layer application.
- *Sequence number*: This field has a length of 32 bits and indicates the sequence number of the segment's first data byte in the overall connection byte stream. Due to the fact that the sequence number is related to a byte count, sequence numbers in contiguous TCP segments are not incremented sequentially.

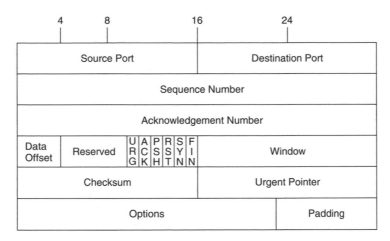

Figure 9.7   The contents of a TCP header

- *Acknowledgement number*: This field has a length of 32 bits and is used by the receiver to acknowledge the receipt of data. It indicates the sequence number of the next byte expected from the receiver.
- *Data offset*: This field has a length of 4 bits and indicates the header length of the segment. It points to the first data byte in this segment.
- *Control flags*: These six flags are used to control several aspects of the TCP connection. The following flags are used:
  Urgent pointer field significant (URG): If this flag is set, then the current segment contains urgent or high-priority data. In this case the value in the urgent pointer field is valid.
  Acknowledgement field significant (ACK): If this flag is set, then the value contained in the acknowledgement number field is valid. With the exception of the first segment, this bit is usually set for the duration of a connection.
  Push function (PSH): If this flag is set, then the transmitting application is allowed to force TCP immediately to transmit the data that is currently buffered without waiting for the buffer to fill. This feature is useful for the transmission of small data units.
  Reset connection (RST): If this flag is set, then the end-to-end TCP connection is immediately terminated.
  Synchronize sequence numbers (SYN): If this flag is set, then the current segment is an initial segment, which is used to establish a connection. This segment contains the initial segment number.
  Finish (FIN): If this flag is set, then the current segment is a request for the normal termination of the TCP connection in the direction in which the segment is travelling. The complete closing of a connection requires one FIN segment in each direction.
- *Window*: This field has a length of 16 bits and is used for flow control. It contains the value of the receiver window size, which is the number of transmitted bytes that the receiver of the segment will accept from the sender.
- *Checksum*: This field has a length of 16 bits and offers basic error detection for the segment.

- *Urgent pointer*: This field has a length of 16 bits and offers the possibility to transmit data which is marked as high-priority by a higher layer application. This urgent data usually bypasses normal TCP buffering and is located in a segment between the header and standard data. The urgent pointer is only valid when the URG flag is set and indicates the position of the first octet of non-expedited data in the segment.
- *Options*: This field has a variable length and is used at connection establishment in order to negotiate a variety of options (e.g. maximum segment size).

9.1.7.2 User Datagram Protocol

UDP offers an end-to-end datagram service. Unlike TCP, it is a connectionless service without the establishment and termination of virtual circuits. Due to this fact, UDP is useful for real-time applications, where the speed is more important than the reliability. The header of a UDP datagram (illustrated in Figure 9.8) consists of the following fields.

- *Source port*: This field has a length of 16 bits and is used to identify the UDP port at the source side of the connection. The use of this field is optional and may be set to 0.
- *Destination port*: This field has a length of 16 bits and defines the destination port at the receiver side.
- *Length*: This field has a length of 16 bits and is used to show the total length of the UDP datagram in bytes.
- *Checksum*: This field has a length of 16 bits and is used for a basic error detection in the UDP datagram.

The above descriptions of the headers used for TCP and UDP clearly show the main advantages and disadvantages of each protocol.

UDP produces much less overhead than TCP, and the connectionless service offers a more effective use of bandwidth. However, non-real-time applications using UDP must realize end-to-end retransmission themselves.

TCP is much more reliable than UDP and can realize node-to-node retransmission. However, TCP produces a large amount of overhead.

9.1.8 A Typical TCP/IP Communication

A typical TCP/IP communication, as illustrated in Figure 9.9, takes place between two hosts connected to the Internet. Starting at the application level on one of the hosts, the following steps are performed:

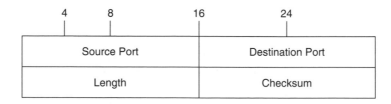

Figure 9.8   The contents of a UDP header

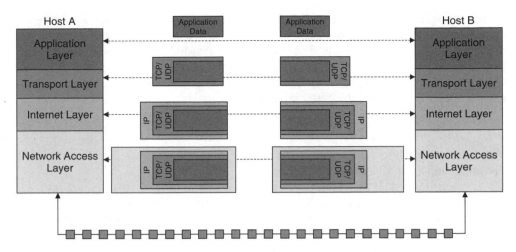

Figure 9.9   A typical TCP/IP communication between two hosts

(1) A network operation is, in general, started at the application level by a user command or by a program. The application sends a continuous stream of information blocks, called messages, to the next lower layer, the transport layer.

(2) The transport layer, where TCP and UDP are implemented, receives the messages from the application. Here, dividing the messages into segments takes place. Each segment gets a transport header and then the segments are passed on to the Internet layer.

(3) The Internet layer, with the related Internet protocol (IP), takes the segments and combines them into an IP datagram and adds the IP header, which contains the destination IP address. The datagram is passed to the lowest layer, the network access layer.

(4) In this layer, the datagrams will be packed in frames (Ethernet) and then transmitted as bit streams over specific network hardware to the destination.

(5) At the destination host, the bit stream is received and reorganized into frames.

(6) Finally, the data goes up through the protocol layers in the reverse direction to the corresponding application.

9.2   Convergence of Fixed, Mobile and Data Networks

9.2.1   The Benefits of Converged Communication

The convergence of fixed, mobile and data networks offers new possibilities for the network operators to make a more efficient use of their networks, and for the end-users, who will profit from new applications and from a more transparent billing system.

The operators can reduce the costs for the network management and for the transmission of data, because now only one network architecture is administered. The use of only one network type requires an effective use of the available bandwidth, but at the same time it offers the opportunity to save money. The combination of the different systems in one solution allows the development of new applications making use of the interaction of voice calls and Internet connections at the same time within the same application.

One example is the possibility of mobile intranet access, which allows employees on business trips to check their e-mails. Additionally, it is possible to enable the employee to have access to all the relevant data and software that he/she requires at the customer site. This example combines mobile and data networks and is a typical situation in which GPRS can give an employee – for example a field engineer – full intranet access from remote sites without using the customer's resources.

9.2.2 Voice-over IP (VoIP)

Voice-over IP (VoIP) is the transmission of speech via packet-switched data networks. The most important goals are:

- the combination of telephone calling capabilities with IP-based networks;
- the interconnection of these capabilities with public telephone networks and private voice networks;
- to offer these services with a high quality.

The basic requirement for voice-over IP is that the packets should arrive at the destination in the correct order, the arrival of all the packets is less important. The realization of VoIP is quite difficult due to a number of problems arising from the different requirements of speech and data transmission. These problems are:

- *large total delay of the packets*: the critical problems that are the result of high end-to-end delay in a voice network are echo and talker overlap. A round-trip delay of more than 50 milliseconds will result in an echo problem. To solve this problem, VoIP systems must implement a kind of echo cancellation. The problem of talker overlap is significant in the case of a one-way delay greater than 250 milliseconds.
- *delay variability (jitter)*: jitter can be explained as the variable inter-arrival times of packets at the destination due to a variable transmission delay over the network. To avoid jitter it is necessary to collect packets and hold them in a buffer in order to allow the slowest packet to arrive in time to be played in the correct sequence.
- *the possibility of packet loss*: IP is not able to guarantee the delivery of packets at all, much less in the correct order. Therefore it is possible that packets will be dropped (lost or discarded) under certain circumstances (peak loads, congestion periods due to link failures or inadequate capacity). The real-time sensitivity of voice transmission doesn't allow the use of TCP-based transmission, but there are solutions that can compensate for the packet loss. Examples are the interpolation of speech by replaying the last packet and the sending of redundant information. Nevertheless, packet losses greater than 10 % are not acceptable.

Despite all these difficulties and limitations, network operators are still interested in solving these problems in order to introduce VoIP. The reason for this is that there are a number of advantages for the operators in transmitting voice over packet data networks. These advantages are:

- *More efficient use of resources*. In a typical phone call, most of the assigned resources are not used, because normally only one partner is talking while the other is listening.

In a packet switched network, bandwidth is only assigned when there is data to be transmitted. Additionally it is possible to reduce the total amount of speech data to be transmitted by using more effective speech compression algorithms.

- *Reduce costs and investments.* A network operator offering speech transmission via the packet-oriented network does not have not to establish a separate circuit switched network.

9.2.3 Multimedia

One of the applications with high potential in the mobile multimedia market is in the field of car navigation and information systems. Such information systems are extremely useful as they enable the driver to optimize his/her route planning without using maps and travel guides. The origin and destination points are entered along with preferences such as fastest or shortest route, and the application does the planning. During a journey, it is possible to follow the route on an interactive map and to receive spoken instructions.

The first systems were CD-ROM based, which, in addition to electronic maps, also stored a large amount of information about hotels, restaurants and other POI (points of interest). Current systems utilize DVDs as storage media due to the fact that the capacity of a DVD is much higher than that of a CD. The main disadvantage of these solutions is that the offered information becomes out of date very quickly. Whenever a CD with the newest information is available, there will soon be some additional information that must be offered to customers as an update. Furthermore, it is not possible to provide information on pricing and availability as these factors change constantly. The introduction of a packet oriented mobile communication system with a large bandwidth like GPRS will offer fascinating new possibilities. It will enable the customers to download the newest information when and where it is required. Additionally, the 'always online' functionality of GPRS can be used to send automatic updates to the customer whenever there are changes, e.g. a traffic jam, road works, or accidents on a route.

Internet, fax, and e-mail are now available in the car, which will enable business users to have a truly mobile office. The introduction of real time protocols will enable video conferences to take place between business users while en route. All of the above will benefit dramatically from the introduction of EDGE combined with GPRS (E-GPRS) which will enable very high speed packet data transfer (see Chapter 5 for more details).

9.3   The Roles of GSM, GPRS and UMTS in Converged Networks

9.3.1   GSM

GSM is one of the standards of the second generation of mobile telephony, so the transmission technology over the radio interface is digital. GSM has been implemented almost worldwide, therefore international roaming is possible. This standard has a much higher capacity than the standards of the first generation. Digital transmission meant that mobile data applications were easily introduced, but not widely used because:

- the transmission of theses applications is not resource efficient as GSM is a purely circuit switched network, i.e. every connection must be set up, maintained with signalling and cleared down at the end;
- the data rates are very low compared with fixed networks;
- only one time slot on the uplink and one on the downlink is available to each user.

9.3.2 GPRS

Data applications have to be transported in a more efficient way, i.e. they must be packet switched. A new packet oriented core network is necessary (SGSNs and GGSNs on an IP network). The GGSNs give a direct interface to the external data networks (intranet, Internet). Also a new distribution of resources over the radio interface has been introduced but in early versions it is not very flexible.

9.3.3 UMTS Release '99

Larger frequency bands of 5 MHz are reserved for the UMTS radio interface for the third generation. Therefore, higher transmission rates are possible, especially when the number of subscribers in a cell is small. Furthermore, the new multiple access principle (code division multiple access CDMA) gives a higher flexibility for the distribution of resources over the radio interface.

A part of the GSM/GPRS network can still be used, but some new components have to be implemented for UMTS:

- User equipment (UE): the UE replaces the MS, the subscriber must purchase a new mobile phone in order to use 3G services. A new SIM card is also required – UMTS uses a USIM.
- Node B: the base station used in UMTS is known as 'Node-B' rather than a BTS. It is able to communicate with CDMA (code division multiple access) subscribers on the new frequency bands, and to support the higher signalling rates used in 3G.
- Radio network controller (RNC): an RNC is similar to a BSC and interfaces with the GSM core network (MSC) for circuit switched calls and the GPRS core network (SGSN) for packet data transfer.

In a second generation GPRS mobile network, no supplementary components are necessary if the ATM switch in the SGSN has been dimensioned with sufficient capacity. The RNC decides, as does the PCU in the GPRS network, whether to route the call to the MSC (voice) or to the SGSN (packet data). An overview of a UMTS network, showing the existing GSM and GPRS network elements, is given in Figure 9.10.

Besides the interfaces to the GSM and GPRS networks, a number of other new interfaces are defined for the implementation of UMTS services based on the introduction of the new network elements Node B and RNC. These interfaces are described in the following text and illustrated in Figure 9.11. The interfaces which were introduced at the introduction of UMTS are:

- Uu: the radio interface between UE and Node B;
- Iub: between Node B and RNC;
- Iur: between RNC and RNC;
- Iu(CS): between RNC and MSC;
- Iu(PS): between RNC and SGS.

On the interfaces Uu and Iub, both circuit and packet switched applications are transported. However, the Iur and Iu(CS) interfaces only carry circuit switched traffic, and the Iu(PS) interface only carries packet switched traffic.

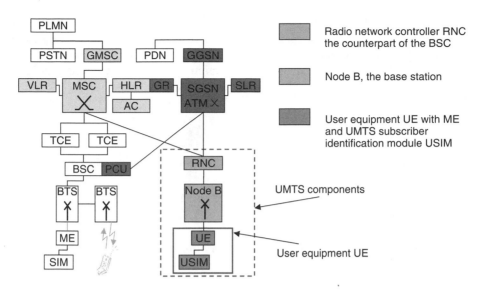

Figure 9.10   New network components for UMTS

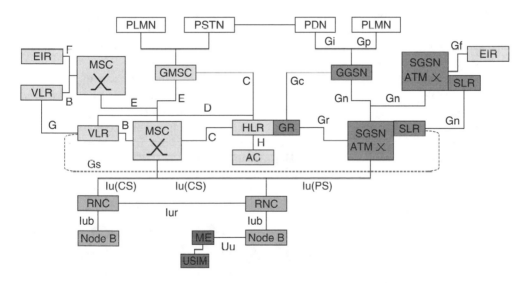

Figure 9.11   The interfaces inside a UMTS network

All these interfaces (except the radio interface Uu) are standardized on ATM for layer 2. ATM (asynchronous transmission mode) is a technology that enables very good QoS to be offered for all possible applications. Therefore data and real-time applications with the corresponding QoS may be transported on these interfaces.

The interface Iu(PS), which only transports packet switched data, is based on IP with the ATM adaptation layer 5 (this adaptation layer is suitable for data applications). The physical layer (layer 1) is usually STM-1 (optical fibre, 155 Mbit s^{-1}).

The interfaces Iub, Iur and Iu(CS) are not IP based. The interfaces Iur and Iu(CS) only transport circuit switched connections, and so make use of the ATM adaptation layer 2, as this adaptation layer is suitable for real-time applications. Layer 1 is usually STM-1 (155 Mbit s^{-1}). The interface Iub transports both circuit and packet switched connections, but only the ATM adaptation layer 2 is standardized. Layer 1 may be E1 (2048 kbit s^{-1}) or STM-1.

9.3.4 UMTS Release 4

In UMTS Release '99, as with GPRS/GSM, the core network is divided into two parts:

- the circuit switched network (all MSC/VLR and GMSC);
- the packet switched network (all SGSN/SLR and GGSN).

Therefore, the network operator has to install, configure and maintain two different networks. The idea with UMTS Release 4 is to merge these two networks into a single entity – this is illustrated in Figure 9.12.

To achieve this we have three options:

(1) To transport everything via the circuit switched network. This was exactly the route followed with the GSM network before the introduction of GPRS. The inflexibility of the resource allocation clearly indicated that this network was not suitable for transporting data applications.
(2) To transport everything on one ATM network. From the theoretical point of view it would be the easiest solution, because ATM is suitable for transporting both real-time and non-real-time applications, but the existing data networks of the operators are usually based on IP and do not always use ATM as the transport medium at layer 1.
(3) To transport everything on the packet switched network, i.e. the IP network. This is the solution proposed by Release 4. Hence, VoIP (voice-over IP) must be implemented here to enable transport of 'packetized' voice traffic.

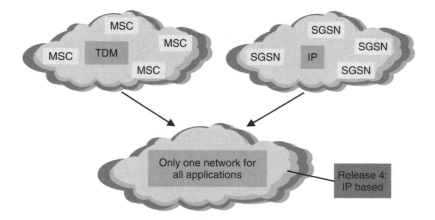

Figure 9.12   The single network concept of UMTS Release 4

The task of transporting real-time applications over IP with a good QoS is certainly more complex than over ATM. But the development of VoIP is still going on, and UMTS Release 4 proposes a solution.

The main change in the network architecture of Release 4 is the separation of the tasks in the circuit switched core network, as illustrated in Figure 9.13. The (G)MSC from the GSM network should be divided into two elements:

(1) The (G)MSC server: this element is responsible for the signalling data, i.e. call control (for example call set-up and call release) and mobility control with the help of the VLR.
(2) The media gateway MGW: this is responsible for the user traffic inside the core network and for the conversion of protocols for the radio network subsystem, for the fixed network (PSTN) or for a pre-Release 4 PLMN.

The idea here, is to use other protocols between the media gateways, e.g. ATM or IP. If the chosen protocol is IP, the MGW and the SGSN can be 'reduced' into one element which is responsible for IP traffic and then we only have one network which transports both real-time and non-real-time applications (see Figure 9.14).

The core network in Release 4 consists of two parts: a traffic network based on IP and an SS7 signalling network. There are two options available with regard to 'how' this SS7 network is implemented:

• based on a circuit switched network with PCM30 lines – as in the older releases,
• based on an IP network.

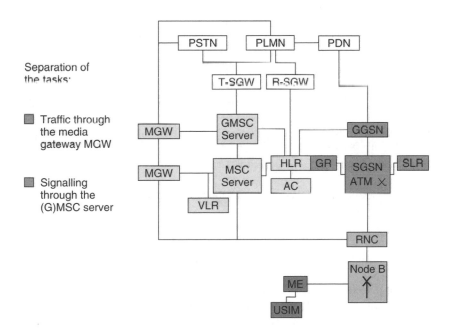

Figure 9.13   The new network elements for UMTS Release 4

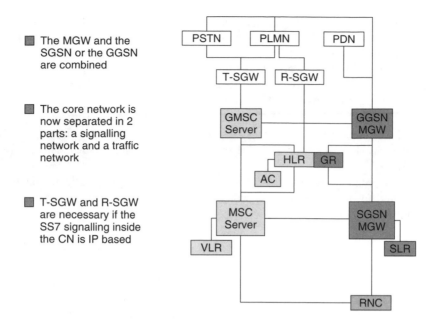

Figure 9.14   The core network over IP in UMTS Release 4

If the second solution is chosen, then two supplementary gateways are necessary:

- the transporting signalling gateway T-SGW: this gateway converts the call-related signalling (e.g. for call set-up and call release) between the PSTN (SS7) or a pre-Release
  4 PLMN (SS7) and the Release 4 network;
- the roaming signalling gateway R-SGW: this gateway performs the signalling conversion (e.g. for roaming, mobility management) between the SS7 based signalling of a
  pre-Release 4 network and the IP based signalling of the Release 4 network.

Some new interfaces (see Figure 9.15) are defined for the implementation of UMTS
Release 4.

The interface Nb is either ATM or IP based. The different layers of each approach are
illustrated in Figure 9.16.

To transport voice-over IP, two protocols are necessary on the transport layer:

(1) Real-time protocol (RTP): this protocol transports real-time applications over a packet
    switched network such as IP.
(2) User datagram protocol (UDP): this protocol is a very simple protocol without any
    acknowledgement, i.e. it is an unreliable protocol. These acknowledgements would
    produce unacceptable delays for real-time applications.

It is also possible to implement the interface Nb on an ATM network.

The interface Mc is based on MEGACO. This protocol addresses the relationship
between the MGW (which converts circuit-switched voice to packet-based traffic) and
the MSC server (which dictates the service logic of that traffic).

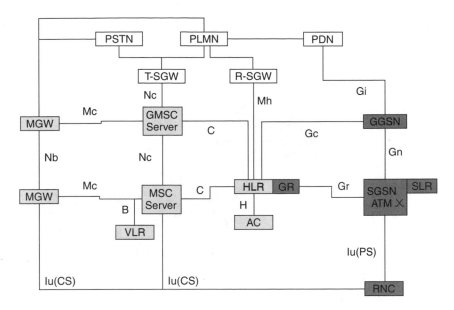

Figure 9.15 New interfaces inside UMTS Release 4

Transport Layer	RTP over UDP	
Network Layer	IP	
Data Link Layer	L2 (like Gn)	AAL2 and ATM
Physical Layer	L1 (like Gn)	L1 (e.g. STM-1)

Figure 9.16 Two different implementations of the Nb interface

The interface Nc is either a classical SS7 interface or is IP based. In the latter case, different protocols may be implemented. For example:

- The bearer independent call control (BICC). This signalling protocol supports narrow-band ISDN services over a broadband backbone network, e.g. ATM or IP.
- SS7 over SCTP/IP. The stream control transmission protocol (SCTP) is necessary for transporting the SS7 messages on an IP interface.

If the signalling network is IP based, the R-SGW is implemented. On the Mh interface, we again have SS7 over SCTP/IP. In which case, all SS7 interfaces C, D, E, F, G and Gc, Gd, Gf, Gr, Gs are also implemented over SCTP/IP.

The interfaces Iu(CS) and Iu(PS) do not change: Iu(CS) is still based on AAL2/ATM/L1 and Iu(PS) on IP/AAL5/ATM/L1 (frequently STM-1 for L1).

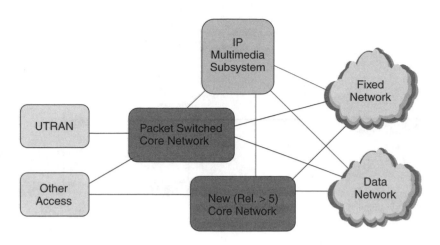

Figure 9.17   IP multimedia subsystem (IMS)

9.3.5   UMTS Release 5

UMTS Release 5 has a control layer that is responsible for handling the signalling for multimedia sessions – this is the so called IP multimedia subsystem (IMS, Figure 9.17). The IMS is an extension of the packet switched core network, and is intended to become independent of the packet switched core network from Release 6 on.

The IMS is only associated with the packet switched core network because in the future voice-over IP might be implemented and it is a prerequisite that all the data of a multimedia service goes through the same core network.

Figure 9.18 illustrates the network architecture of Release 5 with the IMS. For the sake of simplicity, the circuit switched domain is not shown on this diagram though it is included in Release 5. As in Release '99 or Release 4, the circuit switched domain is connected to the UTRAN with the Iu(CS) interface.

The structure of the Release 5 core network is actually quite simple. For the traffic we have the SGSN with the SLR (SGSN location register) database, which serves the subscribers inside the SGSN area, the GGSN, which offers interfaces to external networks (fixed networks and data networks), and the MGW, which connects our IP based network to other circuit oriented networks. For the signalling we have the CSCF, which performs the following session control services for the subscribers:

- session control, origination, termination;
- interaction with services platforms;
- recording of call data.

The CSCF can act as a proxy CSCF (P-CSCF), a serving CSCF (S-CSCF) or as an interrogating CSCF (I-CSCF). The P-CSCF is characterized by being the first contact point for the UE within the IMS. The S-CSCF actually handles the session states in the

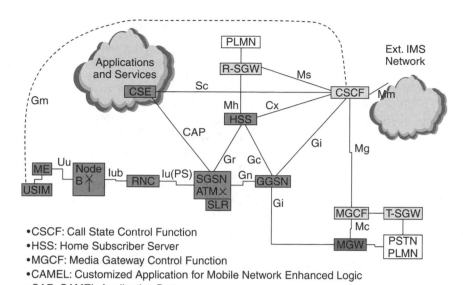

- CSCF: Call State Control Function
- HSS: Home Subscriber Server
- MGCF: Media Gateway Control Function
- CAMEL: Customized Application for Mobile Network Enhanced Logic
- CAP: CAMEL Application Part
- CSE: CAMEL Service Environment

Figure 9.18   IMS domain architecture and interfaces

network, and the I-CSCF is mainly the contact point within an operator's network for all connections intended for a subscriber to that network operator.

Then we have the HSS that substitutes the HLR/AC when the IMS is implemented, the MGCFs that control the MGWs as in Release 4, the MSC servers for the MGWs, and again the T-SGWs and the R-SGWs.

The session initiation protocol (SIP) has been standardized as the call and session control protocol for the IP multimedia subsystem. SIP is a signalling protocol defined by the IETF (Internet Engineering Task Force) for the establishment, release and modification of service sessions in IP networks, such as multimedia conference, multiparty, telephony over IP or chat. SIP is not involved in the transport of session data (the transport can also be in charge of a circuit switched network), but simply carries the information regarding the session (session name, time elapsed since the start, terminal characteristic, transport protocols involved, etc.).

All the interfaces between the elements within the IMS and the UE (which acts as a SIP user agent) are based on SIP, i.e. the interfaces Mg, Mm, Ms, Mw (interface between 2 CSCFs inside one IMS) and Gm (to the UE).

The interface Sc to the services platforms is based on SIP+, which is an enhancement of SIP for service control but is presently not specified.

All the interfaces to the HSS are based on SS7 over SCTP/IP, i.e. the interfaces Gc, Gr, Cx and Mh. If the signalling is not IP based, the interfaces Gc, Gr, Cx are based on the 'classical' SS7 and the Mh interface is not necessary.

As in Release 4, the interface Mc between a MGW and its corresponding MGCF is based on MEGACO.

The traffic interfaces Gi, Gn, Gp inside the core network are based on IP with flexible solutions for the layers 1 and 2.

For the implementation of the interfaces Iu(PS), Iu(CS) and Iub we have two options:

- ATM transport option: This option is the 'old' option. Iu(PS) is based on IP/AAL5/ATM with flexible solutions on layer 1 (usually STM-1). Once again, we may note that the interface Iu(CS) is no longer relevant if real-time applications are transported over the packet switched domain. If this is not the case, Iu(CS) is based on AAL2/ATM with flexible solutions on layer 1 (usually STM-1). Finally, Iub is based on AAL2/ATM with flexible solutions on layer 1 (usually E1 or STM-1).
- IP transport option: This option is the 'new' option. Iu(PS) is based on IP with flexible solutions on layer 2 (i.e. frame relay or Ethernet also possible) and layer 1. Once again, we may note that the interface Iu(CS) is no longer relevant if real-time applications are transported over the packet switched domain. If this is not the case, Iu(CS) is based on RTP/UDP/IP with flexible solutions on layer 2 and layer 1, like the interface Nb in Release 4. Finally Iub is based on IP with flexible solutions on layer 2 and layer 1.

10

Applications

The increase in data rates and efficient use of resources make GPRS attractive for any data-intensive mobile application, especially if it is not sensitive to variations in delay and order of packet delivery. This chapter will explore various applications of GPRS.

10.1 Services

10.1.1 WAP – Wireless Application Protocol

10.1.1.1 Overview

During the last ten years, there has been a significant growth in the mobile telecommunications market and also the Internet market. Today, there are around one billion mobile phone users and about 500 million Internet users. The obvious next step, therefore, was to combine these two rapidly growing markets. The wireless application protocol (WAP) is the link between the mobile telecommunications development and the development of Internet services. It was developed by the WAP Forum, the standardization organization for the mobile Internet, in order to offer a uniform standard for mobile Internet services. The WAP Forum, which is today a part of the Open Mobile Alliance (OMA), was founded in 1997 by four companies Ericsson, Motorola, Nokia, and Unwired Planet (today 'Openwave') Half a year later the WAP Forum was opened up to other companies and today more than 500 companies related to the mobile communications market are now members of the WAP Forum. Figure 10.1 shows the evolution of the WAP standard.

The WAP protocol suite enables mobile network operators to offer a large number of Internet services optimized for mobile use. These services can be divided into several groups:

- *Internet applications*: This category includes information services (such as telephone directories, Yellow Pages, news, and weather forecasts), entertainment services (such as online gaming, multimedia messaging and video streaming applications), and interactive services (such as booking of hotels, flights and car reservations).
- *Banking and financial applications*: This category includes payment procedures via the mobile account, online transactions (e.g. stock exchange order) and access to bank accounts.

GPRS Networks.   G. Sanders, L. Thorens, M. Reisky, O. Rulik, S. Deylitz
© 2003 John Wiley & Sons, Ltd.   ISBN: 0-470-85317-4

• June 1997	Ericsson, Motorola, Nokia and Unwired Planet found the WAP Forum
• September 1997	The basic WAP architecture is published
• January 1998	The WAP Forum is opened for all interested companies
• April 1998	Release WAPv1.0 is published
• May 1999	Release WAPv1.1 is published
• December 1999	Release WAPv1.2 is published
• June 2000	Release WAPv1.2.1 is published
• August 2001	Release WAPv2.0 is published

Figure 10.1   The evolution of the WAP standard

- *Location dependent services*: This category includes services offering information related to the current position of a mobile phone user. Examples are traffic information, local weather forecasts and information about hotels, restaurants and any other POIs (points of interest).
- *Business applications*: This category includes services that enable employees to access their account inside the company network via a mobile network. Examples are access to e-mail accounts, calendars, and other relevant data necessary for the daily work of an employee.

10.1.1.2   The WAP Architecture

The basic architecture of WAP is very similar to the fixed-network Internet based on the TCP/IP protocol stack. In both cases, there is a client–server architecture, where the client sends a request to a server, which sends the requested content in a response. The main difference in the WAP scenario is the client, which in this case is a mobile device. Due to this fact, the mobile Internet solution has some serious limitations compared with Internet access via a fixed network.

- *Limited bandwidth*: In GSM systems a maximum data rate of 9.6 kbit s^{-1} is available over the air interface. Compared with the ISDN data rate of 64 kbit s^{-1} in fixed networks, this 9.6 kbit s^{-1} is not enough for transmitting web pages written in the HTML (hypertext mark-up language) description language for Internet content.
- *Limited CPU and RAM*: The limited RAM doesn't allow the storage of large data files and the limited CPU doesn't allow programs such as Java Scripts to be run, as they require a high processor capacity.
- *Limited devices*: The mobile devices offer only a small display that can only handle a few lines of text or low resolution graphics. These displays are not useful for

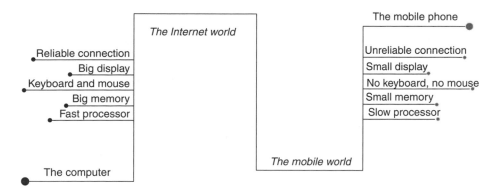

Figure 10.2   The limitations of mobile internet

sophisticated graphics, pictures and coloured animations. In most cases, only text-based information is offered. Another limitation of the mobile device is the keyboard, which doesn't allow the comfortable input of text. Therefore, interactive mobile applications should avoid text input as far as possible.

These limitations (see also Figure 10.2) illustrate the reason why the TCP/IP-based Internet architecture using HTML as the description language cannot be used in the mobile world.

The development of the WAP protocol suite offers a protocol stack that uses the limited resources of the air interface much more efficiently than the TCP/IP protocol suite. The traditional Internet communication uses about 40 % of the total amount of traffic not for payload, but for overhead. The WAP protocol suite produces much less overhead (only about 14 %).

Additionally, a new presentation language for mobile Internet content called WML (wireless mark-up language) and a simplified script programming language for interactive applications (WMLScript) have been developed.

The use of two different protocol stacks requires a new entity that allows a mobile device with WAP capability to start a request to a server located in the Internet. This entity is the WAP gateway, which converts a WAP-based request from a mobile device into a standard HTTP-based request that is used in the Internet world and vice versa (see Figure 10.3). The WAP gateway is the portal for the end-user to the Internet.

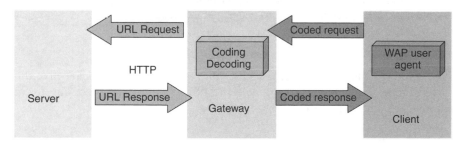

Figure 10.3   The WAP 1.x architecture

WAP 2.0 without Proxy (end-to-end)

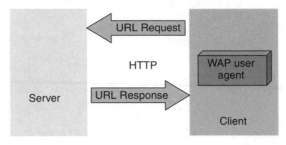

WAP 2.0 with Proxy (split)

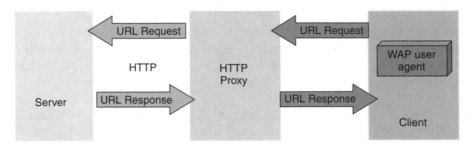

Figure 10.4   The WAP 2.x architecture

The introduction of GPRS increases the bandwidth available via the air interface. As a result of this, the WAP Forum has developed the new WAP release, WAP 2.0, the main task of which is the convergence of the mobile world and the Internet. The release WAP 2.0 brings the wireless world closer to the Internet by supporting all the protocols used for standard Internet applications (IP, TCP, HTTP). With the support of the Internet protocols the WAP 2.0 stack offers the opportunity to distinguish between two architecture models (Figure 10.4).

In the first model, the gateway (as a protocol conversion unit between the mobile client and the origin server) is no longer required. The communication between the mobile device and the server is enabled via HTTP/1.1.

In the second model, the gateway is still present, but here it is used as a proxy and allows the optimization of the communication process and the offering of new services, such as push services or location based services (see Section 10.1.3).

10.1.1.3   The WAP Protocol Suite – WAP 1.x

The WAP protocol suite has a layered architecture and can be divided into five different layers (see Figure 10.5). Each layer is able to fulfil specific tasks according to the OSI reference model. The five layers are:

- wireless datagram protocol (WDP);
- wireless transport layer security (WTLS);

Figure 10.5   The WAP 1.x protocol stack

- wireless transaction protocol (WTP);
- wireless session protocol (WSP);
- wireless application environment (WAE).

The WDP enables the interfacing with the different bearer systems by adapting the transport layer of the underlying bearer. It presents a consistent data format to higher layers of the WAP protocol stack. The main advantage of this architectural concept is the bearer independence of WAP, which offers the possibility of running WAP via different systems such as GSM, GPRS or UMTS.

The WTLS implements security features in the WAP protocol suite, which are based on the standard Internet protocol transaction layer security (TLS). It includes data integrity checks, privacy and authentication checks. The data integrity checks compare the received data with a checksum based on the original data. If someone changes the data during the time the information is en route from the source to the destination, then the checksums of received and sent data are different, and this data packet can be dropped. Privacy means that there is a procedure that allows the communication partners to define a method of exchanging information via a secure channel and no one else can follow their communication. Authentication is the procedure whereby both sides of the communication show their partner their identity. This layer is necessary to enable security sensitive applications like mobile banking, access to company internal data via a mobile device and other e-commerce based applications. For non-critical applications it is not necessary to implement the security layer.

The WTP can run on top of the WDP or on top of the WTLS. It provides a simplified transaction protocol (compared to TCP) which is optimized for low bandwidth scenarios. The WTP offers three classes of transaction services, each class with different bandwidth requirements and reliability properties:

- *Class 0*: corresponds to a one-way, unconfirmed data transport. No confirmations in case of a successful transport are sent and no retransmission of lost data packets is initiated. This transport class requires less bandwidth, but offers no reliability. It can be compared to the user datagram protocol (UDP), which is used as a fast, non-reliable transmission protocol in the Internet.

- *Class 1*: also corresponds to a one-way data transport, but it provides the capability of a confirmed data transport. The transport of data takes place in only one direction and the confirmation (acknowledgement) for the transferred data is transmitted in the opposite direction. This procedure requires more bandwidth but obviously it offers more reliability compared to class 0.
- *Class 2*: provides confirmed transport of data in both directions. If confirmation is not received for a data packet, the sender assumes that it has been lost and starts the retransmission of the lost packet. This system increases the amount of traffic and therefore needs a lot of bandwidth, but at the same time it also provides a high level of reliability. Class 2 is only used for applications where reliability is a prerequisite, such as banking applications.

The WSP is responsible for the coupling of the processes on the client and the server. It offers mechanisms for establishing and releasing such a logical connection and can be compared to the Internet standard HTTP/1.1, including some enhancements such as long-lived sessions and session suspend/resume mechanism. For the transfer of content, WSP requires a transport protocol. WSP offers a consistent interface for two different types of session services. Depending on the UDP/WDP port defined in the profile setting of the phone, this can be a reliable transfer of data based on top of WTP and may also offer a security layer, the WTLS. This type of service is known as a connection-oriented service. The second type of session service can be described as a connection-less service, which allows an unreliable and non-secure data transfer via the WDP directly.

The WAE offers a framework for the different applications and defines the user interface on the mobile device. The applications operate as agents in the WAE application layer of WAP. The most popular application is a micro-browser, which allows navigating through the Internet via mobile networks. The browser is based on WML (wireless mark-up language), which is the language used for presenting content on devices with limited capabilities, and on WMLScript which is the script language that allows the creation of interactive applications.

10.1.1.4   The WAP Protocol Suite – WAP 2.0

Due to the fact, that WAP 1.1 has some limitations (limited size of messages, no real end-to-end security) and the enhancements in the bearer technology that offer a higher bandwidth (HSCSD, GPRS), the WAP Forum has released a new set of specifications for WAP, the WAP 2.0 standard. The main goal of WAP 2.0 is the convergence of mobile Internet and the 'classic' Internet. Therefore, the WAP 2.0 release supports the Internet protocols when IP connectivity is available to the mobile devices (see Figure 10.6). If networks do not provide IP traffic or only low-bandwidth IP bearers can be used, then the WAP 1.x stack is also supported. Both stacks are supported in WAP 2.0 and they can provide similar services to the application environment. The following section describes the new protocols included in the WAP release 2.0.

- wireless profiled hypertext transfer protocol (WP-HTTP);
- wireless profiled transport layer security (WP-TLS);
- wireless profiled transmission control protocol (WP-TCP).

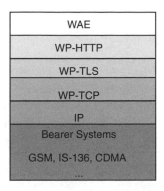

Figure 10.6   The WAP 2.x protocol stack

The WP-HTTP is a derivative of HTTP and is used in a wireless environment. It is fully interoperable with HTTP/1.1 and based on a HTTP request/response transaction between the mobile WAP-capable device and the WAP proxy/server. The main features of WP-HTTP are the support of message body compression of responses and the establishment of real end-to-end secure tunnels.

The WP-TLS allows interoperability for secure transactions. It includes features like cipher suites, certificate formats, signing algorithms and defines the method for TLS tunnelling to support the real end-to-end security at the transport level.

The WP-TCP is fully interoperable with standard Internet TCP implementations and offers optimized performance for a wireless environment. The main features are the support of window sizes larger than 64 kb, the use of large initial windows (up to 4380 bytes) and the path MTU (maximum transfer unit) Discovery, which allows the optimization of the packet size that can be transmitted via the transmission path.

10.1.1.5   Comparison of WAP via GSM-CSD and via GPRS

In the case of WAP being used over GSM circuit switched channels (see Figure 10.7), which is the standard case, the WAP stack in the mobile and the WAP gateway ends with the WTLS layer (if secure transmission is required). The WAP stack is based on the user datagram protocol (UDP). Fragmentation is offered by the network layer, which is realized using the Internet protocol (IP). The connection between the mobile device and the network access server (NAS) is established via point-to-point protocol (PPP) and the connection between the NAS and the WAP gateway is realized via fast Ethernet (100 Mbit s^{-1}).

When comparing this architecture with WAP over GPRS), some important differences must be noted (compare Figure 10.7 with Figure 10.8).

GPRS offers an IP-based backbone network that is used for data traffic, therefore a RAS (which forms the bridge between the circuit-switched and packet-switched network parts) is not needed. The GGSN offers the connection to other networks, such as the internet for example, via the WAP gateway. With the access point name (APN), stored in the GGSN and in the mobile device, it is possible to set up an IP connection to the GGSN. Starting from this point, a connection to the WAP gateway can be established. The IP address of the gateway is also stored in the mobile device. These connections

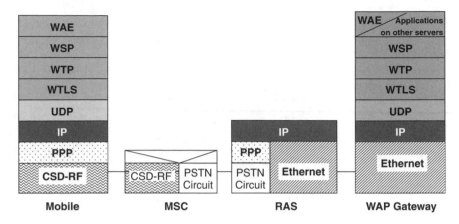

Figure 10.7   WAP over GSM-CSD architecture

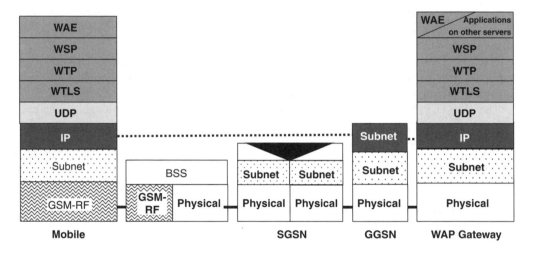

Figure 10.8   WAP over GPRS

form a tunnel within a tunnel, through which the content address in form of an URL is finally sent to the WAP gateway.

10.1.1.6   WAP Applications

The enhancements of the WAP 2.0 release, combined with the advantages of the packet-oriented GPRS system, offer a large number of new applications that may increase the use of mobile Internet. The following section shows some typical examples of how GPRS can be used to optimize the portfolio of mobile data services.

The WAP 'push' functionality is a service that allows a service provider to send (or 'push') content to mobile devices from server-based applications without the need for a user to send a request from the mobile device, i.e. they are subscription based services. The push functionality is especially relevant to real-time applications like messaging, stock price and traffic information. In circuit-switched bearer systems such as GSM, the push messages are transmitted only via the SMS bearer, because this is the only possibility for transmitting current information based on real-time events. With the introduction of GPRS and its 'always online' functionality, it is possible to send push messages via a packet-oriented network. Therefore, providers are no longer limited to SMS for the transmission of push messages.

The multimedia messaging service (MMS) is an application that enables a WAP client to provide a messaging operation with large number of different media types. The main prerequisite for MMS is a large bandwidth for the transmission of a rich set of content, and with the introduction of GPRS this prerequisite has now been met (for more details see Section 10.2).

10.1.1.7 Outlook

The introduction of GPRS offers new perspectives for mobile Internet services. With the advanced technology of GPRS, higher data rates and an effective use of existing resources can be realized that allows the portfolio of existing applications to be enhanced. New business models (volume-based billing, flat rates, etc.) enable the network operator to create attractive pricing systems for different target groups (business men, teenagers, travellers) and therefore to enhance the income-generating traffic via their networks. Intelligent and user-friendly applications will tempt more customers to use these services and more satisfied customers will want to use new services. The development of new services will increase the number of customers and the use of these applications. The result is a win–win situation for customers and operators.

GPRS is the bridge between the circuit-switched GSM system and the packet-oriented systems of the third generation and gives a first impression of how mobile data communication can be realized in the future.

10.1.2   i-mode

10.1.2.1 Overview

i-Mode is a mobile phone platform, which was developed by NTT DoCoMo (Japan) and provides mobile access to e-mail services, Internet content and online gaming, amongst others. It was introduced in Japan in February 1999 and in Germany (e-plus) in March 2002. Attractive business models and content (see Section 10.1.2.3) for content providers and end-users are the reasons for the success of i-mode in Japan. Today, there are more than 30 million i-mode subscribers registered in Japan (Figure 10.9).

Compared with WAP (see Section 10.1.1), the i-mode service has specific features that are responsible for the far greater success of i-mode. i-mode is a proprietary service for mobile data services (Internet, e-mail), that was optimized for the Japanese market. The i-mode service was offered from the beginning as a packet-oriented service, which allows a flexible, attractive pricing model for the end-user. WAP, as a completely bearer-independent protocol suite, was introduced when only circuit-switched services were available in the mobile world. In this case, the billing was time based, and therefore quite expensive for end-users accessing the Internet via mobile networks.

Another difference is the fact, that i-mode was developed as a complete solution for special content services and e-mail services. WAP was developed for mobile Internet access only, and the e-mail service was added later.

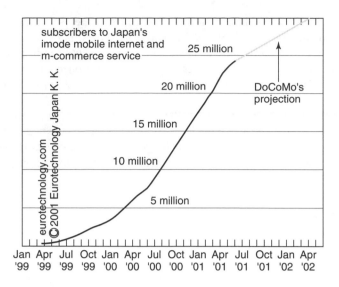

Figure 10.9 i-Mode users in Japan (reproduced with permission from *www. eurotechnology.com*)

10.1.2.2 Architecture

The following components are necessary to offer the i-mode service in a mobile network:

- *A packet data network*: To offer i-mode as packet-oriented service, a packet-based network is necessary. In Japan the PDC (personal digital cellular) standard is used. In Germany the GPRS standard is used as bearer system for the i-mode service.
- *A mobile device with i-mode browser*: A mobile device with i-mode capability is necessary. A large portfolio of mobile devices with different features (coloured display/monochrome display, display size, display resolution, support of sound) is offered in Japan. In Europe, only a few i-mode capable devices are available.
- *An i-mode server (i-mode portal/gateway)*: In analogy to the WAP gateway (see Section 10.1.1), the network operator offers a portal, which allows registered i-mode subscribers to access the i-mode services. The i-mode portal enables licensed content providers (whose content is accessible directly via the portal menu) to be distinguished from unlicensed content providers (whose content is accessible manually via typing the URL).
- *Content providers*: The separation of licensed and non-licensed content providers, which is done by the network operator, allows control of the overall quality of the offered i-mode services via the i-mode portal.

Figure 10.10 shows the whole architecture with all the components mentioned above.

10.1.2.3 Business Model

The basic components of the i-mode business model are: the billing system, which is based on data volume, and the separation in licensed and non-licensed content providers.

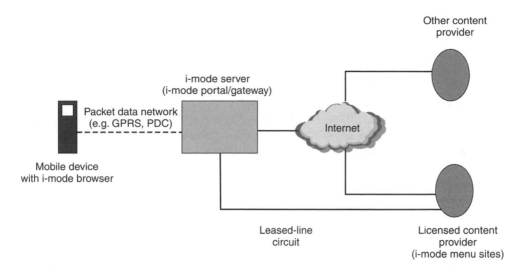

Figure 10.10   The i-mode architecture elements

The volume based billing system is a multi-layered, flexible model consisting of the following components.

- *i-Mode basic fee*: This is a monthly fee for the i-mode service, independent of how often the i-mode service is used and how much data is downloaded.
- *Fees for single services and content*: These are fees for the use of specific services, independent of the amount of data transferred. The content provider is responsible for the pricing model used for its services and content.
- *Fee for the amount of data*: This is the fee for the data transfer (price per kbyte) and it covers the access to i-mode pages and for sending and receiving i-mode mails.
- *i-Mode mail*: Additional to the fee for the amount of data, there is an additional fee for each i-mode mail which is sent

10.1.2.4   i-mode Services

The i-mode contents that are offered in Europe and Japan are different, but there are also some similarities. The most popular content types are news, stock quotes and entertainment applications, e.g. downloading screensavers and ringing tones. The following list shows the i-mode services offered by the German i-mode provider:

- news and weather,
- sports,
- chat, mail, and web,
- entertainment and fun,
- melodies and pictures,
- travel services,
- financial services,
- shopping,
- dictionary services.

10.1.2.5   Outlook

Mobile data services, like i-mode, that can be offered via a packet-based bearer system offer the network operator and the end-user a large number of advantages. In this case, the introduction of GPRS provides a chance for the i-mode service to export the success from Japan to other countries. The German i-mode service (e-plus) via GPRS is the first test of the acceptance of i-mode outside of Japan.

Besides volume-based billing, which offers the network operator the possibility of providing an attractive and flexible billing system for their customers, GPRS technology offers additional features that enhance the capabilities of the i-mode services.

- A large bandwidth combined with the bundling of up to eight channels, allows the download of large data files in a short time.
- The 'always online' capability enables the end-user to receive i-mode mails without a noticeable time delay.

The combination of an attractive and useful service with an intelligent and flexible bearer system such as GPRS is the key for the success of i-mode in the future. There is only one limitation. Due to the fact that i-mode is a proprietary system, there could be interoperability problems with other solutions.

10.1.3   Location Based Services

10.1.3.1   Overview

Location based services are services provided by a mobile network operator or a third party that offer information related to the geographical position of a mobile device user. Studies have shown that out of the whole range of possible mobile data applications, location based services offer some of the greatest potential. Not only can mobile commerce and advertising be realized as a location based service, but also a large number of value-added services such as Yellow Pages, travel assistance, or tracking services. Together with large bandwidth systems like GPRS and adequate mobile devices, these services can be offered in a more user-friendly way (high resolution graphics, coloured maps, etc.). Such services are discussed in the following section.

10.1.3.2   Location Based Services – Categories and Examples

The large number of available and future services can be divided into five broad categories including all the individual services:

- safety services,
- information services,
- tracking services,
- remote services,
- billing services.

Safety services are services requested in an emergency case or for special recovery assistance purposes, e.g. roadside breakdown assistance. This kind of service can be requested manually by pressing an emergency button on the device to initiate a emergency call, or it can be started automatically by on-board systems when a certain event occurs, e.g. the air bag system in car is deployed.

The *information services* include traffic information, Yellow Pages, weather forecasts, specific POI (points of interest) and car parking information systems. The offered content can be static or dynamic, and can be delivered to a large variety of mobile devices. In addition to the most popular mobile phones also smart phones, PDAs, vehicle navigation and information systems, and laptops can make use of these services. Information services can be enhanced by personalization and by use of broadband, packet-oriented bearer systems like GPRS, which allow the offering of push and pull services, and the transmission of large content files such as high resolution coloured maps, animations, etc.

Tracking services take information about the cell in which a subscriber is located, and transform it into a geographical position that can then be overlaid on an electronic map. These services can be used to monitor/find vehicles, cargo, packages, and also people. As such they are useful to business users for such things as fleet and delivery management, and to private individuals for such things as finding their friends in a city and for protecting their children. As an alternative to the cell identity, the GPS (Global Positioning System) could also be used as GPS equipped mobiles are already available on the market.

Remote services are activated from a remote station, usually from a central control station, which activates a process on a mobile unit. Examples are the locking or unlocking of a vehicle or building via a remote control centre or the remote calibration of a truck's engine in order to optimize efficiency.

Billing services offer special pricing models according to the location of the customer. This feature allows mobile operators to compete with wire-line operators. With GPRS it is also possible to offer attractive pricing models for mobile data services, e.g. Internet access via a laptop, based on the location of a subscriber.

10.1.3.3 Architectural Elements

Location based services and implementation details are defined in a set of standardization documents published by ETSI and 3GPP (GSM 03.71, TS 23.271). The functionality provided by the following entities is necessary to make location based services available for circuit-switched networks and for packet-switched networks.

- serving mobile location centre (SMLC);
- location measurement unit (LMU);
- gateway mobile location centre (GMLC);
- home location register (HLR), home subscriber server (HSS).

The SMLC is responsible for the overall coordination and scheduling of resources that are necessary to perform the positioning of a mobile device in a PLMN. The PLMN controls a number of LMUs (location measurement units) in order to obtain the radio interface measurements used to locate a mobile device (or to support other location methods). The SMLC receives the location requests via its associated BSCs and, after the final location estimate and accuracy calculation, the SMLC returns this information in a 'location response' message to the requesting BSC. There can be more than one SMLC in one PLMN.

The LMU is responsible for the radio measurements that can be used by different positioning methods and offers the data to the associated SMLC. The measurements can be divided up into two categories. The radio measurements can be specific to one mobile device and will be used for the location determination of this subscriber (location measurements), or the measurements can be specific to all mobile devices located in a certain geographic area (assistance measurements).

There are two types of local measurement unit (see Figure 10.11)

- *LMU type A*: An LMU type A is connected like a 'normal' mobile station via the air interface on a dedicated SDCCH (stand-alone dedicated control channel). It has its own unique IMSI and its own subscription profile in an HLR. A network operator may assign specific ranges of IMSI for its LMUs and may assign certain digits within the IMSI to indicate the associated SMLC. Certain digits in the IMSI may also be used as a local identifier for an LMU within an SMLC.
- *LMU type B*: An LMU type B, which is basically an additional card inserted in the BTS, is accessed over the Abis interface from a BSC.

The GMLC is the first node an external client accesses in a GSM PLMN. The GMLC stores the location service subscription information on a per LCS (location service)-client basis. When the GMLC receives an LCS request, it uses this information to identify the requesting LCS client and to authorize the LCS client to use the specified request. The LCS client is an application which makes use of location information. The GMLC requests information from the HLR of the mobile device that is being located. The subscriber's privacy information is then used by the GMLC to verify that the LCS client is allowed to locate the mobile subscriber. The GMLC also receives the final location estimates and determines whether they satisfy the requested QoS (quality of service) for the purpose of retry/reject. It can also transform a received location estimate to geographic coordinates

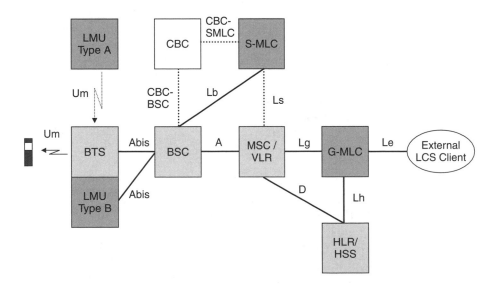

Figure 10.11   ETSI standard LCS architecture

and forward this information to the requesting LCS client. Additionally, the GMLC is responsible for LCS related charging and billing.

The HLR/HSS contains the LCS subscription data and routing information, and is accessible from the GMLC. It is also possible that the HLR/HSS stores the location privacy settings of the subscriber and information about the LMUs (location measurement unit).

10.1.3.4 Positioning Methods

The basis for all location based services is the information about the location of a subscriber. Several methods for retrieving the location information of a mobile subscriber have been standardized. The following are the most important methods:

- Cell ID and timing advance (TA),
- Time of arrival (TOA),
- Enhance observed time difference (E-OTD),
- Assisted global positioning system (A-GPS).

The accuracy provided by the cell ID is limited. Only in big cities with small sectorized cells can the cell ID information be considered sufficiently accurate for use by location based services. In order to enhance the capabilities of the cell ID method, the timing advance (TA) method can be combined with the cell ID information. This approach uses the existing timing advance parameter, which depends on the distance of the MS from the BTS. This value is known by the serving BTS during a call. If the mobile device is in idle mode, special signalling (which is not noticed by the GSM subscriber) is used in order to obtain TA values. TA is used both to assist all positioning mechanisms and as a fall-back procedure. The cell ID in combination with TA is a useful positioning method for applications (weather forecasts, traffic reports) that do not require a high level of accuracy. All this is illustrated in Figure 10.12. in which we can see that the cell identity limits the area in which an MS can be (especially if it is a sector cell), while TA defines a curve on which the MS can be found.

The TOA (time of arrival) method was the first improvement made to the cell ID/TA method. In this scenario, the propagation time of signals exchanged by the mobile device and at least three LMUs is measured. The position information is available in geographic coordinates and has an achievable accuracy of up to 150 metres. The accuracy is determined by the number of LMUs.

The enhanced observed time difference (E-OTD) method uses the same basic idea as TOA (Figure 10.13). It also measures the propagation time of certain signals on the air interface. The difference is that the mobile device is now enhanced with special software, which reduces the number of LMUs to be deployed by a factor of three. However, now the mobile device must see at least three base transceiver stations (BTS) for positioning, which limits the use of E-OTD to urban/suburban areas. In the countryside it is not possible to fulfil this requirement. An accuracy better than 100 metres can be achieved.

The Global Positioning System (GPS) provides the highest accuracy for location information. GPS uses satellites emitting radio signals to a receiver, which can be combined with a mobile device, in order to determine the position of this receiver on the earth's surface. A basic requirement for the high accuracy of GPS is the use of at least three satellites. The positioning measurement of the GPS receiver is based on the TOA principle.

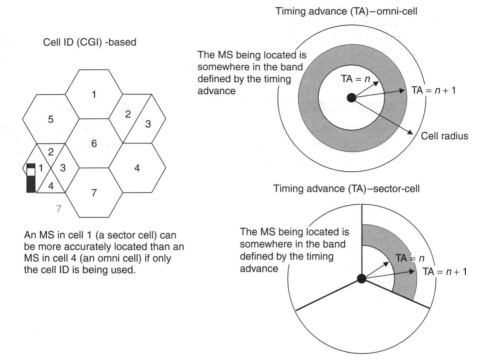

Cell ID (CGI) -based

An MS in cell 1 (a sector cell) can be more accurately located than an MS in cell 4 (an omni cell) if only the cell ID is being used.

Timing advance (TA) – omni-cell

The MS being located is somewhere in the band defined by the timing advance

TA = n

TA = n + 1

Cell radius

Timing advance (TA) – sector-cell

The MS being located is somewhere in the band defined by the timing advance

TA = n

TA = n + 1

Figure 10.12   Positioning via cell ID and TA

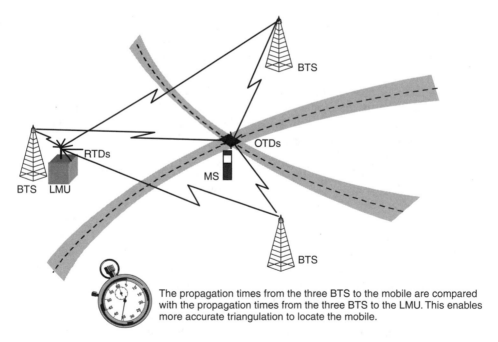

BTS

OTDs

RTDs

BTS   LMU

MS

BTS

The propagation times from the three BTS to the mobile are compared with the propagation times from the three BTS to the LMU. This enables more accurate triangulation to locate the mobile.

Figure 10.13   Positioning via E-OTD

The use of four or more satellites that are in line of sight from the receiver (or receiving antenna) allows the determination of the latitude, longitude and the altitude of the receiver. The current GPS architecture consists of 24 satellites orbiting at an altitude of around 20 000 km above the surface of the Earth. Assisted GPS offers some enhancements. The basic idea is to establish a GPS reference network (or a wide-area differential GPS network) whose receivers have clear views of the sky and can operate continuously. This reference network is also connected with the GSM network. At the request of an MS or network-based application, the assistance data from the reference network is transmitted to the MS to increase performance of the GPS sensor. When the position is calculated at the network then it is called a mobile assisted solution. In the case of position calculation at the handset, it is called a mobile based solution. A properly implemented assisted-GPS method should be able to (Figure 10.14):

- reduce the sensor start-up time,
- increase the sensor sensitivity, and
- consume less handset power than conventional GPS does.

Table 10.1 shows a summary of the main features of the discussed positioning methods.

10.1.3.5 Outlook

The high potential of location based services in the mobile application sector combined, with the enhancements of GPRS, offers the possibility of successful introduction of location based services in packet-oriented networks. The larger bandwidth of GPRS enables the application providers to offer more content in a more user-friendly way. The always-online capability of GPRS allows the use of tracking services very effectively. The possibility

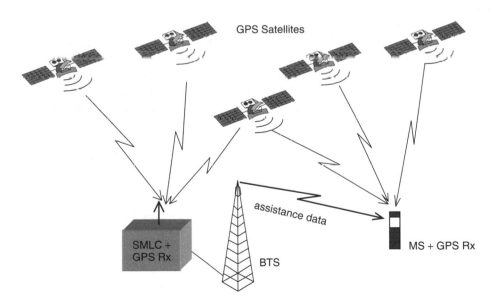

Figure 10.14 Positioning via assisted GPS

Table 10.1   Positioning methods

	Cell ID/TA	TOA	E-OTD	A-GPS
Accuracy	Low	High	High	Very high
Impacts on mobiles	Supported by all mobiles	Supported by all mobiles	New E-OTD capable mobiles required	New mobiles with GPS receiver required
Network modifications	Low (optional LMU, optional SMLC)	High (LMU per BTS, SMLC)	Medium (fewer LMUs than in TOA, SMLC)	Medium (SMLC)
Additional signalling traffic	Very low	High	Medium/low	Medium/low

of volume-based billing offers an additional advantage for the end-user, due to the fact, that he/she can study the information, e.g. map, without paying for the time. The use of location based services via GPRS can also be seen as a pilot project for UMTS. Success stories, problems and unsuccessful projects in this area can be used as helpful examples for the introduction of such services in a UMTS based network.

10.2   Multimedia Messaging Service (MMS)

10.2.1   Overview

The standardized multimedia messaging service (MMS) is a system application that allows a mobile client to provide a messaging operation with a variety of media types. MMS offers a rich set of content to mobile subscribers in the form of a messaging context and supports sending and receiving MMS messages. With MMS it is possible to transmit not only text, but also graphics, sound, and video sequences. At the moment, the most attractive application is the sending of picture postcards. In this case it is possible to send pictures (some mobile devices offer an integrated digital camera) via the mobile network. Other possible applications are high-quality info services, like animated weather forecasts or traveller information. Figure 10.15 shows an example of a multimedia traveller service.

Normally, a non-real-time delivery system is used, but also a real-time messaging solution exists, i.e. instant messaging.

10.2.2   Architectural Elements

A basic multimedia message system consists of the following entities:

- *MMS client*: The MMS client is a service endpoint located on the WAP-capable mobile device.
- *MMS proxy relay*: The MMS proxy relay offers access to a large number of different messaging systems. (It may operate as a WAP origin server in which case it may able to utilize features of the WAP system.)
- *MMS server*: The MMS server is the core component of an MMS system and provides storage services and operational support for the MMS service.

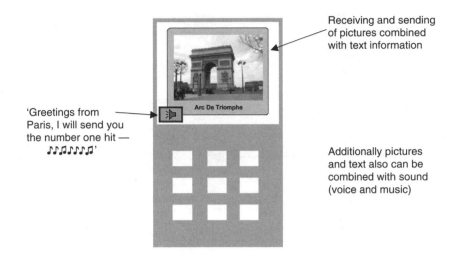

Figure 10.15   Example of a multimedia messaging service (MMS)

There are also a number of additional components that can be used to enhance the capabilities of an MMS architecture:

- *E-mail server*: This server provides traditional Internet email services.
- *Legacy wireless messaging services*: This represents the currently existing systems, e.g. SMS and paging system.

In order to fulfil the requirements of an MMS architecture the following interfaces are defined:

- MMS_M: This is the interface between the MMS Client and the MMS proxy relay.
- MMS_S: This is the interface between the MMS proxy relay and the MMS server. It can be an internal interface in the case of the proxy relay and the server being on the same physical machine.
- MMS_R: This is the interface between the MMS proxy relays of separate MMS systems. Currently there is no standard interface defined.
- *E*: This is the interface between the MMS proxy relay and Internet-based e-mail systems, e.g. SMTP or POP. At the moment this interface is not standardized and only proprietary solutions are available.
- *L*: This is the interface between the MMS proxy relay and legacy wireless messaging systems. This interface is not yet standardized.

Figure 10.16 shows the architecture of an MMS system.

10.2.3   Outlook

The MMS has the capability of being as successful as the SMS, which is the most important data service at the moment. MMS offers the flexibility of SMS, but with many more features. With the introduction of GPRS as a bearer with a large bandwidth, it is

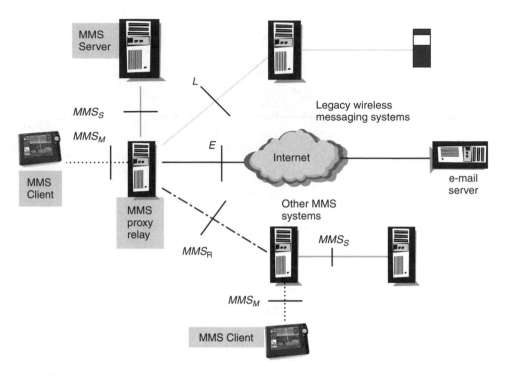

Figure 10.16   Architectural elements of a multimedia messaging service (MMS)

possible to realize the full potential of MMS. Additionally, the multimedia messaging service is the first step in the direction of UMTS and with a successful introduction of MMS via GPRS networks the acceptance of data services for UMTS will also increase.

10.3   GSM-R

European railways are introducing a new communications standard based on GSM. At the time of writing there are 32 railway operators in 24 countries who have agreed to implement the new system. Some of these, such as Banverket in Sweden, already have an operational system covering over 3000 km of track, other countries have test tracks in operation, while others are still in the planning phase.

GSM-R is designed to replace older analogue radio and train control systems using GSM to send and receive both voice and data. Data intensive applications here include:

- automatic train control,
- passenger information systems,
- passenger service systems,
- on-board fault logging and diagnostic systems,
- ordering systems for on-train catering services,
- telemetry.

Such systems require a secure reliable data link between the trains, the network and centralized servers. The consequences of a network failure in GSM-R, in the worst case scenario, could be an accident resulting in loss of life. Safety aspects therefore have absolute priority, which means the standard GSM has had to be refined to conform to very strict requirements in this area. For example: call set-up time, handover and probability of coverage, must be within very strict limits, 2 seconds, 99.5 % success rate, 95 % probability, respectively.

As we have already seen, with GPRS there are no 'calls' to set up or 'handovers' to make, but in most cases the limitations imposed by GSM-R directly impact GPRS. For example, when the mobile is responsible for choosing the most appropriate cell to receive from, probability of coverage is particularly important.

10.3.1 Migration to GPRS

The foundations for the smooth migration to GPRS have already been laid. GSM-R equipment manufacturers have included GPRS functionality in their mobiles, cab radios and base stations, in anticipation of the deployment of the packet data service. If the radio side is already prepared for GPRS, the following are the main changes that must be made to implement it:

- new software in the BTS to control channel coding and Um measurements;
- new hardware and software in the BSC to implement the PCU;
- new packet oriented network elements: SGSN and GGSN;
- new IP network to realize the Gn interface linking the new network elements;
- new interfaces: Gb (PCU to SGSN) and signalling links to HLR, EIR, CSE, etc.

In addition to putting the network infrastructure in place, the operator must decide:

- whether to use dedicated or flexible allocation of GSM/GPRS channels;
- which entity has control of BTS selection – the MS or the PCU;
- what kind of power control to implement for the uplink;
- which data is to be carried via GPRS and which should remain circuit switched;
- what fail safe precautions need to be taken.

The first three points will be determined by such aspects as: which release of GPRS is being implemented, cell planning, the dimensioning of the network, anticipated traffic volumes, safety aspects, etc.

With regard to the potential impact of network element failure consideration must be given to redundancy.

- In the cell planning, BTSs along a route are divided between two BSCs. Consecutive cells are not dependent on the same BSC, so if one BSC fails the network remains operational. Some operators are implementing 'double coverage', which means that the base station planning is done in such a way that any point along the track is covered by two independent BTS. For this reason PCU-controlled cell reselection might be favoured.
- In terms of capacity, one MSC is enough for a national GSM-R network but at least two should be implemented to ensure service if one fails. In Germany, for example, it is planned to use seven. Similar considerations will have to be made for SGSNs.

- As we saw in the implementation section, the easiest way to implement GPRS is to site the first SGSN at an MSC and use nailed up connections (NUC) to realize the Gb interface. Again it is sensible to site an SGSN at both MSCs to reduce the risk of complete loss of service.

Safety and associated risk will also determine which applications will be first to utilize GPRS. The following examples illustrate this point and are presented in the probable order in which they will be realized:

- Passenger services and freight tracking would be low risk, potentially revenue generating, applications that have no train safety aspects.
- Telemetry is a field with a broad scope, which may have a high impact on train safety and therefore carry higher associated risk.
- Automatic train control applications carry the highest risk as train safety depends directly on fast, reliable data transfer.

Each of the above is covered in more detail below.

10.3.2 Automatic Train Control

Train-control applications generally stream small data packets at irregular intervals, typically 50 to 500 bytes at intervals from 2 to 250 seconds or more. GSM-R weighs the advantages and disadvantages of maintaining a CS connection between an on-board computer and control centre for the entire journey against those of re-establishing a PO connection for each batch of data. The GPRS alternative raises the issue of 'time-critical' applications. As with all packet-oriented systems, GPRS cannot guarantee the exact time of arrival for a particular packet. This is because packets from the same source can go via different routes to the destination and hence have differing transfer times, i.e. they can arrive out of sequence and must be buffered and sorted. However, maximum transmission times can be estimated accurately if the system is properly designed. A self-contained network can be dimensioned to optimize packet transfer times, so that data can be transmitted virtually in real time. As the GSM-R networks will be privately owned, built and operated by the railway operator, such dimensioning should be feasible.

Tests need to be conducted on live GSM-R networks to provide reliable information on actual performance, but even if it is not ideal for train control, GPRS is certainly viable for non-time-critical control applications and value-added services.

10.3.3 Passenger Services

GPRS makes a range of information services possible – both on the train and in the station – that are aimed at increasing customer (passenger) satisfaction.

For example, display screens in the train can provide information about onward connections as the train approaches each station. The information is online and current, and can include such things as: platform numbers, platform changes, departure times, delays and other last-minute data. Similarly, such information could be accessed interactively from terminals on the train along with added features such as online ticket purchasing and seat reservations. The terminals would access a central server, via GPRS, to provide up-to-date information on availability. Such terminals could also be situated in stations and function in the same way, i.e. using GPRS for access to the network rather than being hardwired to a LAN.

10.3.4 Freight Tracking Services

Freight tracking is a typical GPRS application. GSM modules combining GPS satellite receivers and GPRS packet data transfer are already available. Such modules attached to containers pinpoint the location of the containers. At regular intervals, this information is gathered and analysed by tracking software. The cargo company knows the exact whereabouts of its shipments in real time, and it can forward the data to its clients and undertake special measures if there is a delivery deadline problem. Freight status data is also available, for example, to enable the cargo company to monitor the temperature of its refrigerated containers. When the module reports that a critical temperature has been reached, the company can alert the freight centre so that technicians can take action when the train arrives at the next station.

Existing commercial systems use the standard GSM frequencies and therefore generate revenue for the PLMN operators. Such systems could easily be implemented using the GSM-R infrastructure and could therefore be billed by the railway operators, i.e. freight tracking could be an additional source of income for the railways.

10.3.5 Telemetry

The GSM modules described above can also be used in rail switching. In this application, the modules would automatically report the position of the rail switching point to the traffic control centre.

The list of possible applications is practically endless. Railway operators could also use GSM modules to gather data from tanker fluid-level indicators, monitor how the loads are secured, report on rail crossings, etc.

The benefit of using GPRS enabled GSM modules lies in the fact that they do not have the same call set-up requirements as GSM and don't block the network with 'fixed connections'. This is because each operates in the standby mode described previously and 'goes active' to send small amounts of data either at regular intervals or as and when required. They are also relatively inexpensive, easy to install and very reliable.

Systems using GSM modules are very limited in their input capability. For more complex applications, on-board systems could use the cab radio to transmit data via GPRS. For example, a network of sensors can be deployed to check train integrity (i.e. that all wagons are still attached) en route. This essential function is currently carried out by expensive trackside train detection equipment.

Another promising application is train diagnostics. Onboard networks monitor the train's status and detect irregularities. The train can send this information via GPRS to the yards in advance, or as the train rolls into the station. The system alerts technicians, who can then be ready and waiting to fix any problems. Minor technical defects are prevented from becoming major or critical, and in the event of serious problems a new locomotive can be standing by. This reduces down time and increases the availability of locomotives, unit trains, and other rolling stock. The customer also benefits from reduced delays and fewer cancellations.

A final example of the flexibility which GPRS brings to GSM-R is surveillance. The high data rates make it possible to send images from cameras mounted on station platforms, shunting yards and car parks to a central security office. Using GPRS-enabled GSM modules linked to security cameras eliminates the need for extensive cable networks, i.e. the cameras can be installed wherever they are required, all they need is a power source and a transmitter.

10.3.6 Driving Forces

There are many driving forces pushing towards the implementation of GPRS in GSM-R networks. The spectrum allocated to GSM-R is small – two blocks of 4 MHz each – which translates as 19 UL channels and 19 DL channels. Conventional CS data transfer via GSM requires that a voice channel is set up and reserved exclusively for one 'data call', regardless of whether the volume of data to be transferred is measured in bits or megabytes. Radio resources are rapidly depleted when many sources transmit small data packages simultaneously in a given sector, especially when the number of channels on the air interface is limited, e.g. a GSM-R BTS will typically have only two carriers, which means 15 available time slots. Further, when many small data quantities are to be transferred, set-up time ends up being a large percentage of total network usage.

As we have already seen, GPRS – with channel bundling, high data rates, fast channel acquisition and release, time slot sharing, and 'connectionless' transmission – offers a flexible, realistic and economically viable solution to all the above.

10.4   m-Business and m-Commerce

10.4.1 m-Business

With the introduction of GPRS and the related mobile devices the role of mobile business applications becomes more important. m-Business applications can be separated into the following groups:

- field services,
- fleet management,
- sales force management and support,
- mobile office.

The *field services* applications will enable employees (especially the service people) of a company to have direct access to the company network (LAN) in order to obtain latest drivers, firmware updates, or patches for certain products. Such access will remove the need for:

- having large amounts of software on a hard-drive;
- carrying around software libraries/archives on CD-ROM;
- making use of a client's telephone or LAN for downloading required software.

Of course such mobile access to a company intranet or server raises important security issues but also reduces the risk of proprietary software being lost by (or stolen from) a company field service engineer.

The *fleet management* applications are based on the use of location based services (LBS) and the technology behind them. Fleet management is the commercial use of such technology in order to administer car or truck fleets. Two examples are tracking services and pick-up services.

With tracking services, the driver of a company vehicle is given a company mobile that is LBS activated. The company administration has a software application that translates the location information from the mobile into coordinates that can be plotted on an electronic map. The administration therefore, has an overview of the current position of all company vehicles at any time. This form of monitoring ensures that all vehicles are used only for company business and also enables assistance to be sent quickly to any vehicle in case of an accident or breakdown. Customers also benefit from such information services as they can be kept informed of the current positions of goods in transit, along with estimated delivery times and news of any problems on route.

The use of the above tracking also enables goods to be picked up from customers by the nearest vehicle. Previously a dispatcher would receive a request from a customer for goods to be picked up for transport and, in the best case, would radio various drivers to see who was the closest to the customer. In the worst case scenario, there would be no communication possible and the customer would have to wait. Using the tracking service outlined above, the dispatcher can quickly identify the nearest vehicle and call the driver to divert to the customer to pick up the goods. In the best case, the driver will have a PDA, which can log all the details of the customer and the goods to be transported. This information will be sent from the driver to the central administration server via GPRS, rather than via an exchange of faxes or e-mails that would then require manual entry of the data into the system.

The tracking systems described above are already available and use location information based on cell identity. Future systems are likely to use GPS coordinates from GPS receivers built into the mobiles, as this will give much better accuracy.

The *sales force management* applications are used for the administration of mobile sales staff and interconnection via company LAN, which can be enhanced with all the mentioned functions such as database access via virtual private data and speech networks. Information from sales personnel falls into three main categories: queries about availability or stock, estimated delivery times, and orders.

Much of this information was previously exchanged via voice calls, which meant sales personnel had to call a central office. This office, therefore, represented a bottleneck in the sales support process. Also, voice communication by telephone is prone to creating errors. The exchange of data (enabled by GPRS) between a mobile device and a central server removes both the bottleneck and reduces the possibility of human errors.

The exchange of such information results in better organization from the central office, easy reorganization of stock, and just-in-time delivery. An example of the above is company sales representatives who are responsible for on-the-shelf stock levels in supermarkets. These reps have a barcode reader with which they scan the product label and then enter the required quantity to be ordered. There are already mobile phones with barcode readers built in. It will be a small step to enable such devices to use GPRS to transmit the data.

The *mobile office* solution is an integration of several different applications and solutions in order to offer mobile employees the same office applications as their colleagues located directly in the office. Such applications will initially be access to e-mail, fax, time planning applications and company intranet. A further possibility will be the actualization of time planning applications, such as Microsoft Outlook ©, between mobile employees and a central server.

A prerequisite for all these m-business applications is a mobile-oriented packet data network with a large bandwidth. GPRS as a packet-oriented bearer system for data transmission is, at the time of writing, the best solution for introducing such mobile business services.

10.4.2 m-Commerce

m-Commerce applications are based on transactions performed via mobile networks. They can be divided into several groups. At the time of writing, the most popular applications that could profit from the GPRS technology, are mobile related selling, mobile based selling, and mobile gaming.

Mobile related selling includes the offering of mobile related 'gimmicks' and 'gadgets' that can be differentiated into soft products and hard products.

Soft products can be delivered in electronic form and include: ringing tones, logos, and SMS-extras (such as bit-map graphics and melodies). The ordering can be done via WAP, SMS or a premium-rate call, and delivery can be via WAP, or enhanced SMS. As these are low cost items, the retailer depends on volume and repeat sales, and the customer can be charged directly via his or her phone bill to make ordering easier and facilitate 'impulse buying'. Here, both WAP applications and SMS can be transported via GPRS and thus benefit from the higher data rates it offers.

Mobile phone related hard-products are those that require physical delivery to the customer and thus include such items: as custom cases, custom covers, stands, SIM cards, interface cables, software on CD-ROM, etc.

Mobile based selling refers to the mobile device (phone, PDA, notebook) being used as a direct sales channel to the customer in the same way as a store or catalogue, and can also be divided into soft products and hard products.

Here, the soft products are those that the customer would like to get because of a personal or professional interest. Examples are 'services' (such as mobile banking applications, brokerage services, push services, location based services,) and 'Internet content' (such as tourist information, city guides, public transport timetables, entertainment news, hotel guides, etc.).

Again, hard products are those that require physical delivery to the customer and thus include almost any saleable product. As such, any product that has been successfully sold via the Internet could be sold in this way. Additionally, retailers could offer display windows in which goods are offered to customers at a discount if ordered online (i.e. via Internet using their mobile), thus motivating the customer to use m-commerce. Airlines are using a similar tactic to reduce costs by encouraging customers to book flights via the internet, i.e. customers need to be motivated to make use of new channels such as the internet and their mobile device.

Mobile gaming applications can be divided into single-player games and network-based multi-player games. The latter especially will profit from the high data rates that are offered by mobile network architectures of the next generation. With current phones, the Java micro edition enables their gaming flexibility but future mobiles, and especially PDAs, will have enough graphics and processing power for online gaming. In the beginning, the gaming server could take over some of the computing tasks from the mobile phone to lower its CPU load. This is, for example already done in WAP environments for certain applications.

For the customer m-commerce offers flexibility and convenience, in that buying decisions can be acted upon immediately, i.e. anywhere and anytime, without having to find the right store or wait for it to open. For the retailer, m-commerce offers an additional channel to the customer and maximizes the potential for impulse buying, i.e. the channel is always available and does not suffer from such limitations as 'opening hours' or 'lack of stock'.

11

Roaming and GRX

11.1 Introduction

With the expansion of GSM networks around the world, the concept of 'roaming' is one that most users now take for granted. 'Roaming' refers to the freedom to use your mobile phone while travelling in other countries that also have GSM networks. Of course there are some prerequisites which enable this freedom:

- your home PLMN (HPLMN) operator must have a roaming agreement with the operator in the country you are visiting;
- your mobile phone must be compatible with the GSM network you are visiting, i.e. mobile and PLMN must both be either GSM900, 1800 or 1900;
- roaming must be permitted by your user licence, i.e. the subscriber data stored in the HLR must indicate that roaming is allowed.

Roaming in GSM is more straight forward than roaming in GPRS. This is because GSM deals with circuit switched connections and billing for call duration. With GPRS, however, the concept of quality of service is introduced with regard to available data rates. Further, the billing is more complex as, ideally, billing should be based on the volume of data transferred.

This chapter will explain how roaming can be realized in GPRS and, in particular, the benefits available from the use of so called GPRS roaming exchanges (GRX).

11.2 Why do we need Roaming in GPRS?

There are many reasons for enabling subscribers to roam into other networks for GPRS services. These include:

- local Internet access in other countries;
- access to e-mail servers;
- access to company intranets.

GPRS Networks. G. Sanders, L. Thorens, M. Reisky, O. Rulik, S. Deylitz
© 2003 John Wiley & Sons, Ltd. ISBN: 0-470-85317-4

The first point is the simplest scenario. The local SGSN only needs to access your HPLMN to authenticate you and obtain your subscriber data. The SGSN can then provide access to the Internet via a local GGSN.

The second point may be slightly more complex. In the case of a global e-mail service, such as Hotmail © or Yahoo ©, the local DNS will quickly be able to locate the correct server via the Internet. However, in the case of country-specific e-mail servers, the DNS query may have to pass through a hierarchy of DNS before the address of the destination server is obtained.

This is also the case with the third point. Imagine the only access point to a certain intranet (e.g. a corporate network) only exists in one specific GPRS network. As long as the user registers at an SGSN of this network, no roaming is necessary. When the user registers in an SGSN of another network the SGSN needs to establish a connection to the GGSN in the user's home network. Therefore, the DNS request will travel outside the boarders of the network. The request from the local DNS will move up through increasingly higher level DNS until one is found that knows which DNS to contact in your home network. The request will then move down through lower level DNS in your home network until the correct address is found for your company intranet.

None of the above is possible unless:

- the operators concerned have roaming agreements with each other;
- the DNS are capable of resolving the addresses;
- there is a connection between the two networks – either directly or via a third network.

11.3 Architecture

In its simplest form, the above requires the following architecture (Figure 11.1):

- DNS
- Border gateways (BGW)
- IP based transport network between the two GPRS core networks

The disadvantage of this approach is that every operator must make a roaming agreement and set up a connection to every other operator, in order to enable roaming in each other's networks. This results in a fully meshed scenario as illustrated in Figure 11.2.

To avoid the disadvantages inherent in the above, the concept of a GPRS roaming network was evolved. This means that instead of establishing multiple links with other operators, each operator establishes a single link to the roaming network, and thus all operators can be connected with each other via this central network. Such roaming networks

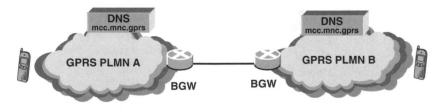

Figure 11.1 Basic architecture necessary for roaming

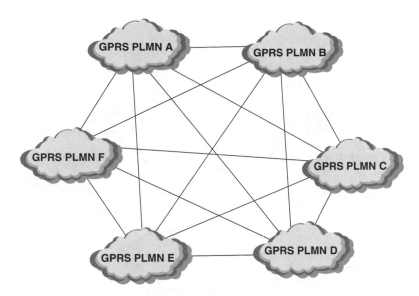

Figure 11.2   Fully meshed scenario

can be established on a regional basis, and these regional roaming networks can also be connected together to expand further the roaming possibilities offered to the subscribers of each individual operator.

11.4   GPRS Roaming eXchange (GRX) Network

The principle configuration for a GRX network is illustrated in Figure 11.3. Here we can see some of the main strengths of a GRX network:

- avoiding the complexity of meshed networks;
- resources can be shared between all connected PLMNs;
- gaining easy access to the central (root) DNS;
- fast implementation of roaming services;
- easy to maintain (one link instead of many);
- unique agreements throughout the GRX and all connected PLMNs (e.g. for connections with specific quality of service parameters).

The advantages of such a configuration are obvious. Anything that reduces complexity, shortens the time-to-market, and enables ease of maintenance, has to be good for both the operator and the subscriber. Further benefits emerge in terms of billing. A billing model can be implemented in which the roaming exchange takes care of billing-record collection and financial clearing. This ensures a transparent process for all parties concerned and simplifies invoicing and billing between the operators.

The GRX concept can be expanded to include 3G and WLAN (wireless LAN), thus enabling roaming to be implemented quickly for these emerging technologies.

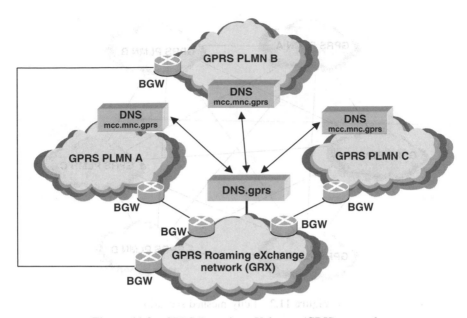

Figure 11.3   GPRS Roaming eXchange (GRX) network

11.5   Procedures

This section explains the basic steps involved in enabling GPRS access for a roaming mobile (Figure 11.4).

In the first example, the MS registers in an SGSN that does not belong to its home network. The intranet that the MS wants to connect to is only accessible via the home network of the subscriber. Therefore, the following roaming procedure is necessary to enable a connection establishment between the two networks via the GRX.

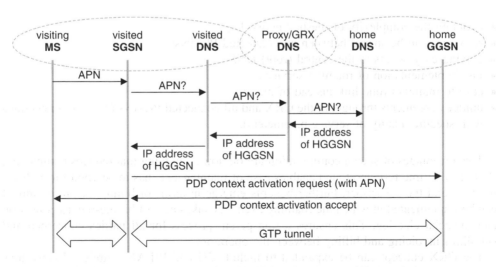

Figure 11.4   Procedure for initiating GPRS access while roaming

(1) The MS requests GPRS access and submits the access point name (APN) of the required server.
(2) The visited SGSN interrogates the local DNS with the requested APN.
(3) The local DNS does not recognize the APN and so interrogates a higher level proxy DNS, or the root GRX DNS if the network is connected to a roaming exchange network.
(4) If the proxy/GRX DNS knows the address it will return it to the visited DNS, otherwise it will contact the DNS in the home network.
(5) The visited SGSN receives the IP address of the required GGSN.
(6) The visited SGSN sends a PDP context activation request to the GGSN.
(7) The GGSN accepts the request and a GTP tunnel is established.
(8) Two-way data transfer is now possible between the mobile and the destination via the GGSN.

This procedure would also be used, for example, to make use of a WAP service in the home network.

In the second example a mobile again registers with an SGSN that does not belong to its home network, but in this case the user only wants to have Internet access. As the Figure 11.5 shows, the procedure is now much simpler as the local DNS is able to obtain the required address without interrogating the home DNS.

When the MS first registers with the network, the SGSN contacts the HLR/GR in the home network to obtain the subscriber data for the user. In this example, therefore, there is no need to contact the home network again and the local DNS can resolve the required internet address. The procedure is as follows:

(1) The MS requests GPRS access and submits the access point name (APN) of the required server.
(2) The visited SGSN interrogates the local DNS with the requested APN.
(3) The local DNS returns the address of the required GGSN.

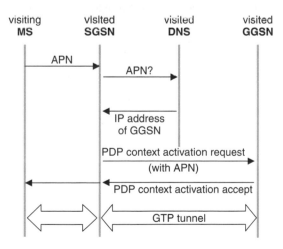

Figure 11.5 GPRS roaming, making use of VGGSN for internet access

(4) The visited SGSN sends a PDP context activation request to the GGSN.
(5) The GGSN accepts the request and a GTP tunnel is established.
(6) Two-way data transfer is now possible between the mobile and the destination via the GGSN.

11.6 Quality Aspects of GRX

When several operators are connected via a single GRX the quality of service (QoS) is provided by this GRX so QoS while roaming can be guaranteed and areas of responsibility can be clearly defined.

When two GRXs agree to peer, then QoS is now provided by two networks. In such cases areas of responsibility can still be relatively clearly defined so it should still be possible to guarantee QoS. The problem arises when three or more GRX are connected. In such cases, areas of responsibility are unclear and QoS cannot be guaranteed.

As a result, GRX must offer the following:

- interconnection between GPRS networks;
- interconnection between other GRX;
- routing of traffic between participating GPRS networks and GRXs;
- a centralized DNS service;
- defined minimum quality of service levels;
- security.

The benefits to the operators are as follows:

- meshed scenarios are not required;
- access to a central DNS;
- maintaining QoS and security are less of a burden.

The use of GRX promises major advantages to operators in quickly establishing global roaming opportunities for their subscribers, reducing start-up costs and maintaining good quality of service levels.

Glossary and Abbreviations

2G	Second Generation
3G	Third Generation
3GPP	Third Generation Partnership Project – a collaboration agreement that brings together a number of telecommunications standards bodies. The current organizational partners are ARIB, CWTS, ETSI, T1, TTA and TTC
4PSK	4 Phase Shift Keying – equivalent to QPSK
8PSK	8 Phase Shift Keying – the modulation method used by EDGE

A

A3	Authentication Key Algorithm – used by the AC and the SIM to generate the SRES (signed response) key (inputs to A3 are Ki and RAND)
A5	Encryption Algorithm – uses the ciphering key (Kc) to cipher a bit sequence
A8	Ciphering Key Algorithm – used by the AC and the SIM to generate the ciphering key, Kc (inputs to A8 are Ki and RAND)
AAL	ATM Adaptation Layer
AB	Access Burst
ABR	Available Bit Rate
ACR	Available Cell Rate
AC	Authentication Centre
ACC	Automatic Congestion Control
ACCH	Associated Control Channel
ACIR	Adjacent Channel Interference Ratio
ACK	Acknowledgement
ACLR	Adjacent Channel Leakage Power Ratio
ACM	Address Complete Message
ACS	Adjacent Cell Selectivity
ADC	Analog to Digital Converter

GPRS Networks.   G. Sanders, L. Thorens, M. Reisky, O. Rulik, S. Deylitz
© 2003 John Wiley & Sons, Ltd.   ISBN: 0-470-85317-4

ADN	Abbreviated Dialling Number
ADPCM	Adaptive Differential Pulse Code Modulation
ADSL	Asymmetrical Digital Subscriber Line
AE	Application Entity
AEC	Acoustic Echo Control
AEF	Additional Elementary Functions
AESA	ATM End System Address
AGCH	Access Grant Channel – channel by which the BTS informs the MS which SDCCH to use for authentication, etc.
AI	Acquisition Indicator
AICH	Acquisition Indication Channel
AID	Application Identifier
AIUR	Air Interface User Rate
AK	Anonymity Key
ALCAP	Access Link Control Application Part
ALSI	Application Level Subscriber Identity
AM	Acknowledged Mode
AM	Amplitude Modulation
AMF	Authentication Management Field
AMPS	Advanced Mobile Phone Service – the analogue 1G cellular mobile phone network very popular in the USA
AMR	Adaptive Multi-Rate – method for changing the coding scheme used for voice calls to suit the changing quality of the air interface
AN	Access Network
AoC	Advice of Charge
AP	Access Preamble
APDU	Application Protocol Data Unit
API	Application Programming Interface
APN	Access Point Name
ARFCN	Absolute Radio Frequency Carrier Number
ARIB	Association of Radio Industries and Businesses – Japanese standardization committee
ARP	Address Resolution Protocol
ARQ	Automatic Repeat Request
AS	Access Stratum
ASC	Access Service Class
ASCI	Advanced Speech Call Items – enables shared downlink group calls
ASE	Application Service Element
ASN.1	Abstract Syntax Notation One
ATDM	Asynchronous Time Division Multiplexing
ATM	Asynchronous Transfer Mode
ATR	Answer To Reset
ATT (flag)	Attach
AuC	Authentication Centre

AUTN	Authentication Token
AWGN	Additive White Gaussian Noise
B	
BA	BCCH Allocation
backhaul	routing of a call from the access network to the switching element (which is often geographically far removed from both the origin and destination of the call)
BAIC	Barring of All Incoming Calls
BAOC	Barring of All Outgoing Calls
BAR	BRAIN Access Router
BCC	BTS Colour Code
BCCH	Broadcast Control Channel
BCD	Binary Coded Decimal
BCF	Base Station Control Function
BCFE	Broadcast Control Functional Entity
BCH	Broadcast Channel
BCIE	Bearer Capability Information Element
BEC	Backward Error Correction
BECN	Backward Explicit Congestion Notification
BER	Bit Error Rate
BFI	Bad Frame Indication
BG	Border Gateway
BGT	Block Guard Time
BHCA	Busy Hour Call Attempts
BIC	Baseline Implementation Capabilities
BIC-Roam	Barring of Incoming Calls when Roaming
BID	Binding Identity
B-ISDN	Broadband ISDN
BLER	Block Error Rate
Bm	Full-rate traffic channel
BMC	Broadcast/Multicast Control
BMG	BRAIN Mobility Gateway
BN	Bit Number
BOC	Bell Operating Company
BOIC	Barring of Outgoing International Calls
BOIC-exHC	BOIC except to Home Country
BPSK	Binary Phase Shift Keying
BRAIN	Broadband Radio Access for IP-based Networks
BS	Base Station
BSC	Base Station Controller
BSG	Basic Service Group
BSIC	Base transceiver Station Identity Code
BSIC-NCELL	BSIC of an adjacent cell
BSS	Base Station System
BSSGP	BSS GPRS Protocol – layer 3 of the protocol stack of the interface Gb between the PCU and the SGSN

BSSAP	BSS Application Part – upper layer of the protocol stack of the GSM interface A between the BSS and the MSC (BSSAP is split in two protocols, BSSMAP and DTAP)
BSSMAP	BSS Management Application Part
BSSOMAP	BSS Operation and Maintenance Application Part
BTS	Base Transceiver Station
bundling	the use of more than one time slot by one subscriber, e.g. at the time of publication it was possible to bundle four DL time slots (but only two UL time slots) for one GPRS user
burst	the short radio transmission in which the modulated bits for one time slot are transmitted is known as a radio 'burst', i.e. a BTS transmits one TDMA frame as eight separate bursts on the same frequency
BWT	Block Waiting Time
C	
C	Conditional
C-	Control-
CA	Cell Allocation
CAA	Capacity Allocation Acknowledgement
CAC	Connection Admission Control
CAI	Charge Advice Information
CAMEL	Customized Applications for Mobile network Enhanced Logic
camp	a mobile is said to 'camp on a cell' when it has registered with the network and monitors the CCCH of a particular BTS for paging messages
CAP	CAMEL Application Part
CB	Cell Broadcast
CBC	Cell Broadcast Centre
CBCH	Cell Broadcast Channel
CBMI	Cell Broadcast Message Identifier
CBR	Constant Bit Rate
CBS	Cell Broadcast Service
CC	Call Control
CC	Country Code
CCBS	Completion of Calls to Busy Subscriber
CCCH	Common Control Channel
CCF	Conditional Call Forwarding
CCH	Control Channel
CCITT	Comité Consultatif International Télégraphique et Téléphonique (The International Telegraph and Telephone Consultative Committee)
CCK	Corporate Control Key
CCM	Certificate Configuration Message
CCP	Capability/Configuration Parameter

CCPCH	Common Control Physical Channel
CCPE	Control Channel Protocol Entity
CC/PP	Composite Capability/Preference Profile
Cct	Circuit
CCTrCH	Coded Composite Transport Channel
CCU	Channel Codec Unit
CD	Capacity Deallocation
CD	Collision Detection
CDA	Capacity Deallocation Acknowledgement
CDMA	Code Division Multiple Access – the multiple access used on the radio interface of UMTS
CDR	Charging Data Record (often called a 'ticket') – a record of network use by a subscriber, stored in SGSN (network access) and GGSN (Internet access), which is used for billing purposes
CDUR	Chargeable Duration
CDV	Cell Delay Variation
CDVT	Cell Delay Variation Tolerance
CEIR	Central Equipment Identity Register
cell reselection	the process by which the next cell which a mobile communicates with is selected – can be performed by the MS or the PCU
CEPT	Conférence des administrations Européennes des Postes et Télécommunications
CES	Circuit Emulation Service
CF	Conversion Facility
CFB	Call Forwarding on mobile subscriber Busy
CFN	Connection Frame Number
CFNRc	Call Forwarding on mobile subscriber Not Reachable
CFNRy	Call Forwarding on No Reply
CFU	Call Forwarding Unconditional
CGI	Cell Global Identity – the worldwide unique identifier of a cell: CGI = MMC + MNC + LAI + CI.
channel group	several frame relay virtual channels (VC) multiplexed together. Very efficient as all VCs in the group share the physical bandwidth, i.e. a single VC can take the entire bandwidth when other VCs are not transmitting
CHAP	Challenge Handshake Authentication Protocol
CHP	Charging Point
CHV	Call Holder Verification information
CI	Cell Identity
CID	Charging ID – an ID used to identify the CDRs for one subscriber
CIM	Common Information Model
CIR	Carrier to Interference Ratio

CKSN	Ciphering Key Sequence Number – number associated with the cipher key Kc derived during authentication; with CKSN it is possible for the network to identify Kc stored in the MS without invoking the authentication procedure
CLA	Class
CLI	Calling Line Identity
CLIP	Calling Line Identification Presentation
CLIR	Calling Line Identification Restriction
CLK	Clock
CLNP	Connectionless Network Protocol
CLNS	Connectionless Network Service
CLP	Cell Loss Priority
CLR	Cell Loss Ratio
CM	Connection Management
CMIP	Connection Management Information Protocol
CMISE	Connection Management Information Service
CMM	Channel Mode Modify
CMM	Common Mobility Management – refers to the simultaneous updates of routing area in the SLR and location area in the VLR when a routing area update is performed by a GPRS mobile
CN	Core Network
CNAP	Calling Name Presentation
CNG	Calling tone
CNL	Cooperative Network List
CO	Circuit Oriented
COLI	Connected Line Identity
COLP	Connected Line identification Presentation
COLR	Connected Line identification Restriction
COM	Complete
CONNACK	Connect Acknowledgement
CONS	Connection Oriented Network Service
CORBA	Common Object Request Broker Architecture
CoS	Class of Service
CP	Control Plane
CPE	Customer Premises Equipment
CPICH	Common Pilot Channel
CPCH	Common Packet Channel
CPCS	Common Part Convergence Sublayer
CPS	Common Part Sublayer
CPU	Central Processing Unit
C/R	Command/Response field bit
CRC	Cyclic Redundancy Check (3 bit)
CRE	Call RE-establishment procedure
CRNC	Controlling Radio Network Controller

CS	Coding Scheme
CS	Circuit Switched
CSCF	Call State Control Function
CSD	Circuit Switched Data
CSE	CAMEL Services Environment
CSPDN	Circuit Switched Public Data Network
CT	Call Transfer
CTCH	Common Traffic Channel
CTD	Cell Transfer Delay
CTDMA	Code TDMA
CTR	Common Technical Regulation
CTS	Cordless Telephony System
CUG	Closed User Group
CW	Call Waiting
CWTS	China Wireless Telecommunication Standard group
D	
DAC	Digital to Analog Converter
DAD	Destination Address
DAM	DECT Authentication Module
D-AMPS	AMPS adapted for three time slot digital transmission
datagram	normally refers to a packet of data with a header attached, e.g. an IP datagram is a data packet with an IP header
DB	Dummy Burst
DC	Dedicated Control (SAP)
DCA	Dynamic Channel Allocation
DCCH	Dedicated Control Channel
DCE	Data Circuit terminating Equipment
DCF	Data Communication Function
DCH	Dedicated Channel
DCN	Data Communication Network
DCS1800	Digital Cellular System at 1800 MHz
DDI	Direct Dial-In
DECT	Digital Enhanced Cordless Telephony
dedicated mode	refers to the fact that GSM calls require 'dedicated' resources, i.e. one time slot = one user (on the air interface and in the core network)
DET	Detach
DF	Dedicated File
DHCP	Dynamic Host Configuration Protocol – realized central administration/allocation of an IP address pool
DHO	Diversity Handover
DiffServ	Differentiated Services – a supplementary architecture for offering QoS on the IP layer
DISC	Disconnect

diversity	the use of two antennas in the receive path for better reception – the Rx signal from each is compared and only the strongest is used
DL	downlink, i.e. from BTS to MS
DLCI	Data Link Connection Identifier
DLD	Data Link Discriminator
Dm	Control channel (ISDN terminology applied to mobile service)
DMR	Digital Mobile Radio
DMT	Discrete Multi-Tone modulation
DMTF	Distributed Management Task Force
DN	Destination Network
DNIC	Data Network Identification Code
DNS	Domain Name Server – resolves web addresses, into IP addresses, e.g. a DNS will resolve www.tfk.de into 212.227.119.105
DP	Dial/Dialled Pulse
DPCCH	Dedicated Physical Control Channel
DPCH	Dedicated Physical Channel
DPDCH	Dedicated Physical Data Channel
DRAC	Dynamic Resource Allocation Control
DRNC	Drift Radio Network Controller
DRNS	Drift Radio Network Subsystem
DRX	Discontinuous Reception (mechanism)
DS-CDMA	Direct Sequence CDMA
DSCH	Downlink Shared Channel
DSE	Data Switching Exchange
DSI	Digital Speech Interpolation
DSL	Digital Subscriber Line
DSLAM	Digital Subscriber Line Access Multiplexer
DSS1	Digital Subscriber Signalling No. 1
DTAP	Direct Transfer Application Part
DTCH	Dedicated Traffic Channel
DTE	Data Terminal Equipment
DTMF	Dual Tone Multi-Frequency (signalling)
DTX	Discontinuous Transmission (mechanism)
DWDM	Dense Wavelength Division Multiplex – the multiple access used in optical fibres

E	
EA	External Alarms
EBSG	Elementary Basic Service Group
ECM	Error Correction Mode
Ec/No	Ratio of energy per modulating bit to the noise spectral density
ECSD	Enhanced Circuit Switched Data – the circuit switched part of the EDGE standard
ECT	Explicit Call Transfer

ECTRA	European Committee of Telecommunication Regulatory Affairs
EDC	Error Detection Code byte
EDGE	Enhanced Data rates for GSM Evolution – the new high bit-rate modulation method for Um, using 8PSK
edge router	A term used in ATM networks. It is a device that routes data packets between one or more LANs and an ATM backbone network. Sometimes referred to as a boundary router
EEL	Electric Echo Loss
EF	Elementary File
EFCI	Explicit Forward Congestion Indication
EFR	Enhanced Full Rate – refers to a coding scheme on the GSM air interface
EFS	Error Free Second
E-GGSN	Enhanced GGSN
E-GPRS	Enhanced General Packet Radio Service – the packet switched part of the EDGE standard
E-HLR	Enhanced HLR
EIR	Equipment Identity Register
EIRP	Equivalent Isotropic Radiated Power
EL	Echo Loss
EM	Element Manager
EMC	Electro-Magnetic Compatibility
eMLPP	enhanced Multi–Level Precedence and Pre-emption service
EMMI	Electrical Man–Machine Interface
EPROM	Erasable Programmable Read Only Memory
ERP	Equivalent Radiated Power
ERR	Error
Ethernet	the most widely installed LAN technology. Specified in standard *IEEE 802.3* and comes in four forms: 10BaseT, 100BascT, Gigabit and 10Gigabit, which transmit respectively at 10 Mbit, 100 Mbit, 1 billion bps and 10 billion bps.
ETNS	European Telecommunications Numbering Space
ETR	ETSI Technical Report
ETS	European Telecommunication Standard
ETSI	European Telecommunications Standards Institute
etu	elementary time unit

F

FA	Fax Adaptor
FAC	Final Assembly Code
FACCH	Fast Associated Control Channel
FACH	Forward Access Channel
FAR	Fix Alternative Routing
FAUSCH	Fast Uplink Signalling Channel

FAX	Facsimile
FB	Frequency correction Burst
FBI	Feedback Information
FCCH	Frequency Correction Channel
FCI	File Control Information
FCS	Frame Check Sequence
FDD	Frequency Division Duplex – one frequency for the uplink and another frequency for the downlink
FDM	Frequency Division Multiplex
FDMA	Frequency Division Multiple Access – the method by which many users are given access to a certain part of the radio spectrum by dividing the available bandwidth into separate frequency bands
FDN	Fixed Dialling Number
FEC	Forward Error Correction – the addition of redundancy to a radio block in order to be able to correct transmission errors
FECN	Forward Explicit Congestion Notification
FER	Frame Erasure Rate
FER	Frame Error Rate
FFS	For Further Study
FH	Frequency Hopping – improves security and reduces interference by using a different frequency each time a given time slot is transmitted, i.e. the time slot number remains the same but it is sent on a different frequency in each subsequent TDMA frame (hopping can be cyclic or pseudo-random)
FIFO	First In First Out
FITL	Fibre In The local Loop
FH	Frequency Hopping
FM	Fault Management
FM	Frequency Modulation
FN	Frame Number
FNUR	Fixed Network User Rate
FP	Frame Protocol
FR	Full Rate – refers to a coding scheme on the GSM air interface which compresses digitized speech into 13 kbit s^{-1}
FR	Frame Relay – refers to a form of packet data transmission which is used on the Gb interface between the PCU and the SGSN
fragmentation	a function of IP whereby large packets are 'fragmented' into smaller packets, each of which receives an IP header – not to be confused with the 'segmentation' performed by the SNDCP protocol, in which a complete IP packet is divided into LLC frames.

ftn	forwarded-to number
FTAM	File Transfer Access and Management
FTP	File Transfer Protocol
FTTB	Fibre To The Building/Basement
FTTC	Fibre To The Curb
FTTEx	Fibre To The Exchange
FTTH	Fibre To The Home
FTTL	Fibre To The Loop
FTTN	Fibre To The Node

G

GC	General Control (SAP)
GCR	Group Call Register
GERAN	GSM/EDGE Radio Access Network
GFR	Guaranteed Frame Rate
GGSN	Gateway GPRS Support Node
GID	Group Identifier
GMLC	Gateway Mobile Location Centre
GMM/SM	GPRS Mobility Management and Session Management
GMSC	Gateway MSC
GMSK	Gaussian Minimum Shift Keying – the modulation method used by GSM
GP	Guard Period
GPA	GSM PLMN Area
GPRS	General Packet Radio Service
GPS	Global Positioning System – system that provides the user with current latitude and longitude coordinates derived from triangulation with a minimum of three satellites
GR	GPRS Register – an extension to the HLR in which GPRS specific subscriber information is stored
GR	GPRS Release
GSA	GSM System Area
GSIM	GSM Service Identity Module
GSM	Global System for Mobile communications
GSM-R	special implementation of GSM for Railway operators
GSN	GPRS Support Node – used when making reference to a GPRS node that could be either a SGSN or a GGSN (was also used in early implementations to indicate a combined SGSN/GGSN)
GT	Global Title
GTT	Global Title Translation
GTP	GPRS Tunnelling Protocol
GTP-U	GTP for User Plane
GUI	Graphical User Interface
GW	Gateway

H

HCS	Hierarchical Cell Structure
HDLC	High level Data Link Control
HE-VASP	Home Environment Value Added Service Provider
HF	Human Factors
HHO	Hard Handover
HLC	High Layer Compatibility
HLR	Home Location Register – GSM network element that stores subscriber data and the ID of the current VLR
HN	Home Network
HO	Handover
HOLD	call Hold
hop	As an IP packet moves from the source to the destination, it usually passes through several routers. Each part of the path, from router to router, is known as a 'hop'. Routing decisions are made on a 'hop-by-hop' basis.
HPLMN	Home PLMN
HPS	Handover Path Switching
HPU	Hand Portable Unit
HR	Half Rate – one time slot serves two mobiles by using even numbered TDMA frames for one mobile and odd numbered frames for the other
HRR	Handover Resource Reservation
HSCSD	High Speed Circuit Switched Data – enables a data rate of 14.4 kbit s^{-1} to be used in a time slot on the GSM air interface and also enables one subscriber to make use of several time slots (bundling)
HSN	Hopping Sequence Number
HSS	Home Subscriber Server
HTML	Hyper Text Mark-up Language – the language used to generate a web page
HTTP	Hyper Text Transfer Protocol – the protocol used to transfer a web page to a browser on a PC
HTTPS	HTTP Secure
HU	Home Units

I

I	Information frames
IAM	Initial Address Message
I-Block	Information Block
IC	Integrated Circuit
ICC	Integrated Circuit Card
ICGW	Incoming Call Gateway
ICM	In-Call Modification
ICMP	Internet Control Message Protocol
ICR	Initial Cell Rate

ID	Identification/Identity/Identifier
IDL	Interface Definition Language
IDN	Integrated Digital Network
IE	Information Element
IEC	International Electro technical Commission
IEI	Information Element Identifier
IETF	Internet Engineering Task Force
I-ETS	Interim European Telecommunications Standard
IF	Infrastructure
IFS	Information Field Size
IHOSS	Internet Hosted Octet Stream Service
IK	Integrity Key
IMA	Inverse Multiplexing on ATM
IMEI	International ME Identity – the identifier of the mobile device that may be checked inside the Equipment Identity Register EIR
IMGI	International Mobile Group Identity
IMSI	International Mobile Subscriber Identity
IMS	IP Multimedia Subsystem – a new subsystem to be implemented in the UMTS Release 5
IMT	International Mobile Telecommunications
IMUN	International Mobile User Number
IN	Intelligent Network
INAP	Intelligent Network Application Part
INF	Information Field
IntServ	Integrated Services – a supplementary architecture for offering QoS on the IP layer
I/O	Input/Output
IP	Internet Protocol – the network layer (layer 3) implemented, for example, in the Internet and in the GPRS core network
IP-M	IP Multicast
IPv4	Internet Protocol version 4 – uses 32 bit addressing in the form of four decimal number separated by dots, e.g. 212.227.119.105
IPv6	Internet Protocol version 6 – uses 128 bit addresses in the form of 8 hexadecimal numbers separated by colons, e.g. ABCD:EF54:9854:1023:FDEC:AB31:5363:7733
IR	Infrared
IRP	Integration Reference Point
ISC	International Switching Centre
ISCP	Interference Signal Code Power
ISDN	Integrated Services Digital Network
ISO	International Organization for Standardization
ISP	Internet Service Provider
ISUP	ISDN User Part
ITC	Information Transfer Capability

ITU	International Telecommunication Union
IUI	International USIM Identifier
IWF	Inter-Working Function
IWMSC	Inter-Working MSC
IWU	Inter-Working Unit

J

JAR file	Java Archive file
JD	Joint Detection
Jitter	the variation in the transmission delay of a data packet caused, for example, by packets being sent via different paths to a destination
JNDI	Java Naming Directory Interface
JP	Joint Predistortion
JPEG	Joint Photographic Experts Group
JTAPI	Java Telephony Application Programming Interface

K

kbit/s	kilo-bits per second: $1 \text{ kbit s}^{-1} = 1000$ bits per second (not 1024)
kbps	the same as kbit s^{-1}
ksps	kilo-symbols per second
Kc	Ciphering Key – part of a triple used for ciphering over the radio interface
Ki	individual subscriber authentication Key
killer application	a marketing expression that refers to an application which appeals to a large segment of the population and thereby generates a very large amount of income, e.g. prepaid cards for GSM telephones

L

L1, L2, L3	Layer 1, 2, 3 – the physical layer (1), data link layer (2) and networking layer (3)
LA	Location Area – group of cells within which an MS can move around without having to make a location update
LAC	Location Area Code – the code that forms part of the CGI of a cell and identifies the location area (LA) to which the cell belongs
LAI	Location Area Identity – uniquely identifies a location area within a network (for MM in GSM)
LAN	Local Area Network
LAPB	Link Access Protocol Balanced
LAPDm	Link Access Protocol on the Dm Channel
LATA	Local Access and Transport Area
LAU	Location Area Update
LCD	Low Constrained Delay

LCN	Local Communication Network
LCP	Link Control Protocol
LCS	Location Service
LCSC	LCS Client
LCSS	LCS Server
LDA	Location Dependent Addressing
LE	Local Exchange
LEA	Law Enforcement Agency
LEN	Length
LI	Length Indicator
link	the path followed by an IP packet between two nodes
LLC	Logical Link Control
Lm	traffic channel with capacity lower than a Bm
LMSI	Local MS Identity
LMU	Location Measurement Unit
LN	Logical Name
LND	Last Number Dialled
LNP	Local Number Portability
location update	procedure whereby the mobile recognizes that it has entered a new location area and the ID of the new location area (LAI) is sent to the current VLR – if the new location area belongs to a new VLR then the HLR is also updated with the ID of the new VLR
logical	'logical' entities are described in terms of functionality whereas 'physical' entities are the actual equipment, interfaces and network resources, e.g. the logical interface GGSN-HLR may be physically realized via the SGSN
LPLMN	Local PLMN
LR	Local Register
LSA	Local Service Area
LSAP	Link Service Access Point
LSB	Least Significant Bit
LSR	Load Sharing Routing
LSTR	Listener Side Tone Rating
LT	Line Termination
LTE	Local Terminal Emulator
LTZ	Local Time Zone
LU	Local Units
LU	Location Update
LV	Length and Value
M	
M	Mandatory
MA	Mobile Allocation
MA	Multiple Access

MAC	Media Access Control – a protocol used, for example, to allocate radio resources on the GPRS air interface (a layer 2, MAC protocol is also used to control media access in computer networks, such as an Ethernet)
MAC	Message Authentication Code
MACN	Mobile Allocation Channel Number
MAF	Mobile Additional Function
MAH	Mobile Access Hunting
MAHO	Mobile Assisted Handover
MAI	Mobile Allocation Index
MAIO	MAI Offset
MAP	Mobile Application Part – the application protocol used on signalling interfaces between the MSC and GSM databases, i.e. the: HLR, VLR, AC, and EIR
MBS	Maximum Burst Size
MCC	Mobile Country Code
MCI	Malicious Call Identification
Mcps	Mega-chips per second
MCR	Minimum Cell Rate
MCU	Media Control Unit
MD	Mediation Device
MDS	Multimedia Distribution Service
ME	Maintenance Entity
ME	Mobile Equipment: needs a SIM card in order to be considered a mobile station (MS)
MEF	Maintenance Entity Function
MEGACO	Media Gateway Controller
MEHO	Mobile Evaluated Handover
measurement report	the mobile measures RXLEV and RXQUAL of the BCCH of the current cell and neighbouring cells – the report has many uses including HOV decisions by the BSC and DL power control by the current BTS
MER	Message Error Rate
MExE	Mobile Execution Environment
MF	Master File
MF	Multi-Frame – the logical structure of a physical channel with respect to time, e.g. a GSM TCH has a MF comprising 26 TDMA frames, most of which are for sending speech but some of which are for signalling
MFS	Maximum Frame Size
MGCF	Media Gateway Control Function
MGCP	Media Gateway Control Part
MGT	Mobile Global Title
MGW	Media Gateway
MHEG	Multimedia and Hypermedia information coding Expert Group
MHS	Message Handling System

MIB	Management Information Base
MIC	Mobile Interface Controller
MIM	Management Information Model
MIP	Mobile IP
MIPS	Million Instructions Per Second
MLC	Mobile Location Centre
MM	Mobility Management – procedures that enable the position of the mobile to be accurately maintained in network databases, and updated as and when required
MME	MM Entity
MMI	Man–Machine Interface
MMS	Multimedia Messaging Service
MNC	Mobile Network Code
MNP	Mobile Number Portability
MO	Mobile Originated
MOC	Mobile Originating Call
MOHO	Mobile Originated Handover
MO-LR	Mobile Originated Location Request
MOS	Mean Opinion Score
MoU	Memorandum of Understanding
MPEG	Moving Pictures Experts Group
MPLS	Multi Protocol Label Switching
MPTY	Multi-Party
MRF	Media Resource Function
MS	Mobile Station: MS = ME + SIM
MSB	Most Significant Bit
MSC	Mobile services Switching Centre
MSCM	MS Class Mark
MSCU	MS Control Unit
MSE	MExE Service Environment
MSID	MS Identifier
MSIN	MS Identification Number
MSISDN	Mobile Station International ISDN Number – has the following structure: country code + network destination code + subscriber number CC + NDC + SN (the first two digits of the SN are the HLR-id)
MSN	Multiple Subscriber Number
MSP	Multiple Subscriber Profile
MSRN	MS Roaming Number – enables a GMSC to contact the current MSC of a mobile for the purpose of initiating a paging message in the case of a mobile terminating call (MTC)
MT	Mobile Terminated
MTC	Mobile Terminating Call – a subscriber must be located, paged and authenticated in order to receive a call

MT-LR	Mobile Terminating Location Request
MTM	Mobile-To-Mobile
MTP	Message Transfer Part – collective name for the layer 1,2 and 3 protocols used on signalling interfaces between the MSC and the: BSC, HLR, VLR, AC, EIR, etc.
MTU	Maximum Transfer Unit
MU	Mark-Up
MUI	Mobile User Identifier
MUMS	Multi-User MS
MUX	Multiplexer

N

NAD	Node Address byte
NAI	Network Access Identifier
NAS	Non-Access Stratum
NB	Normal Burst
NBAP	Node B Application Part
NCC	Network Colour Code
NCELL	Neighbouring (of current serving) Cell
NCH	Notification Channel
NCK	Network Control Key
NCP	Network Control Protocol
NDC	Network Destination Code
NDUB	Network Determined User Busy
NE	Network Element
NEF	Network Element Function
NEHO	Network Evaluated Handover
NET	Norme Européenne des Télécommunications
NEV	Never
NF	Network Function
NIC	Network Independent Clocking
NI-LR	Network Induced Location Request
NITZ	Network Identity and Time Zone
NM	Network Manager
NMC	Network Management Centre
NMS	Network Management Subsystem
NMSI	National MS Identification number
NNI	Network-Node Interface
NO	Network Operator
node	element of an IP network (with links to other nodes) responsible for routing decisions and packet relay
NOM	Network Operation Mode: can be I, II or III and indicates how certain procedures are handled, e.g. NOM-I indicates that Gs is implemented and CMM procedures can be used
NP	Network Performance

NPA	Numbering Plan Area
NPI	Number Plan Identifier
NRM	Network Resource Model
NRT	Non-Real Time
NSAP	Network Service Access Point
NSAPI	NSAP Indicator
NSCK	Network Subset Control Key
NSDU	Network Service Data Unit
NSS	Network Switching Subsystem – the circuit switched part of the core network
NT	Network Termination
NT	Non-Transparent
Nt	Notification (SAP)
NUA	Network User Access
NUI	Network User Identification
NUP	National User Part (SS7)
NW	Network

O

O	Optional
O&M	Operation and Maintenance
OAM	Operation, Administration and Maintenance
OACSU	Off Air Call Set Up – relieves busy hour congestion by only allocating a traffic time slot to a MOC when the called party goes off hook
OCCCH	ODMA Common Control Channel
OCF	Open Card Framework
octet	a set of 8 bits in which each bit or group of bits may be interpreted separately, i.e. in an 8 bit byte, all the bits are interpreted together
ODB	Operator Determined Barring
ODCCH	ODMA Dedicated Control Channel
ODCH	ODMA Dedicated Channel
ODMA	Opportunity Driven Multiple Access
ODTCH	ODMA Dedicated Traffic Channel
OMC	Operation and Maintenance Centre
OML	Operation and Maintenance Link
ORACH	ODMA Random Access Channel
OS	Operating System
OSA	Open Service Architecture
OSI	Open System Interconnection
OSI RM	OSI Reference Model
OSP	Octet Stream Protocol
OSPF	Open Shortest Path First – a link state routing protocol used in IP networks

OSS	Operation Sub System
overhead	refers to extra tasks or resources that are necessary to support the processes which generate income – the administration required to support a sales team is overhead, the header(s) on an IP packet is overhead, etc.
OVSF	Orthogonal Variable Spreading Factor

P

PABX	Private Automatic Branch Exchange
PACCH	Packet Associated Control Channel
PAD	Packet Assembly/Disassembly facility
PAGCH	Packet Access Grant Channel
PAP	Password Authentication Protocol
PBCCH	Packet Broadcast Control Channel
PBP	Paging Block Periodicity
PBX	Private Branch Exchange
PC	Personal Computer
PC	Power Control – refers to a control parameter for MS and BTS transmissions
PCB	Protocol Control Byte
PCCC	Parallel Concatenated Convolutional Code
PCCCH	Packet Common Control Channel
PCCH	Paging Control Channel
PCCPCH	Primary Common Control Physical Channel
PCG	Project Coordination Group
PCH	Paging Channel
PCK	Personalization Control Key
PCM	Pulse Code Modulation
PCM30	2 Mbit/s interface consisting of 32 time slots each of 64 kbit s^{-1}. One TS is the frame alignment word so only 31 TS are available for transporting data. Originally another TS was reserved for signalling hence PCM30 (not 31). PCM24 (1.5 Mbit s^{-1}) is widely used in the USA.
PCMCIA	Personal Computer Memory Card International Association
PCPCH	Physical Common Packet Channel
PCR	Peak Cell Rate
PCS	Personal Communication System
PCU	Packet Control Unit
PD	Protocol Discriminator
PDC	Personal Digital Cellular (Japan)
PDCP	Packet Data Convergence Protocol
PDH	Plesiochronous Digital Hierarchy
PDN	Public Data Network
PDP	Packet Data Protocol
PDSCH	Physical Downlink Shared Channel
PDTCH	Packet Data Traffic Channel

PDU	Protocol Data Unit
PG	Process Gain
PH	Packet Handler
PHB	Per Hop Behaviour
PHF	Packet Handler Function
PHI	Packet Handler Interface
PHS	Personal Handyphone System
PHY	Physical layer
PhyCH	Physical Channe0l
physical	*see* logical
PI	Page Indicator
PI	Presentation Indicator
PICH	Page Indication Channel
PICS	Protocol Implementation Conformance Statement
PID	Packet Identification
PIN	Personal Identification Number
PIXT	Protocol Implementation eXtra information for Testing
PLMN	Public Land Mobile Network
PMD	Physical Media Dependent
PN	Pseudo Noise
PNCH	Packet Notification Channel
PNE	Présentation des Normes Européennes
PNP	Private Numbering Plan
POI	Point Of Interconnection (with PSTN)
POTS	Plain Old Telephone System
PP	Point-to-Point
PPCH	Packet Paging Channel
PPE	Primitive Procedure Entity
PPF	Paging Procced Flag
PPM	Parts Per Million
PPP	Point-to-Point Protocol
PPS	Protocol and Parameter Select (response to the ATR)
PRACH	Packet Random Access Channel
PS	Packet Switched
PSC	Primary Synchronization Code
PSCCCH	Physical Shared Channel Control Channel
PSCH	Physical Shared Channel
PSE	Personal Service Environment
PSPDN	Packet Switched Public Data Network
PSTN	Public Switched Telephone Network
PTCCH	Packet Timing advance Control Channel
PTM	Packet Transfer Mode
PTM	Point-To-Multipoint
P-TMSI	Packet TMSI
PTP	Point-To-Point
PU	Payload Unit

PUCT	Price per Unit Currency Table
PUK	PIN Unblocking Key
PUSCH	Physical Uplink Shared Channel
push service	a mobile user subscribes to an information service which regularly sends ('pushes') the requested information across the network to the user
PVC	Permanent Virtual Circuit
PW	Password

Q

QoS	Quality of Service
QPSK	Quadrature (Quaternary) Phase Shift Keying – the modulation method used by UMTS

R

RA	Routing Area – a group of cells within which a GPRS mobile can move around without having to perform a routing area update
RAB	Random Access Burst
RAC	Routing Area Code
RACH	Random Access Channel – the UL channel on which a MS makes first contact with the network (the response comes on the AGCH and directs the MS to a specific SDCCH)
radio block	a radio block is the smallest transmission unit in GSM or GPRS and consists of four time slots which are prepared together using the same coding scheme and transmitted in four consecutive TDMA frames
RADIUS	Remote Authentication Dial-In User Service
RAI	Routing Area Identity – uniquely identifies a routing area within a network (for MM in GPRS)
RAM	Random Access Memory – dynamically addressable temporary storage media
RAN	Radio Access Network – refers to the GERAN for GSM networks and the UTRAN for UMTS networks
RAND	Random Number – basic component of a triple (RAND, SRES, Kc) in which SRES and Kc are based on RAND and a subscribers Ki
RANAP	Radio Access Network Application Part
RAS	Remote Access Server – forms the interface between the CS world of GSM and the IP world of the Internet
RAU	Routing Area Update
RB	Radio Bearer
RBER	Residual Bit Error Rate
RDF	Resource Description Format
RDI	Restricted Digital Information
REC	Recommendation

redundancy	in telecommunications this means having some form of built-in back-up in case something fails, e.g. many racks have $1+1$ redundancy (one back-up module for every active module); all voice and most data transmissions over Um are made with added redundancy (original bits + copied bits)
REJ	Reject(ion)
REL	Release
REQ	Request
RF	Radio Frequency
RFC	Radio Frequency Channel
RFCH	Radio Frequency Channel
RFN	Reduced TDMA Frame Number
RFU	Reserved for Future Use
RIP	Routing Information Protocol – used in IP networks for dynamic updating of routing tables
RL	Radio Link
RLC	Radio Link Control
RLCP	Radio Link Control Protocol
RLP	Radio Link Protocol
RLS	Radio Link Set
RMS	Root Mean Square (value)
RNC	Radio Network Controller
RNS	Radio Network Subsystem
RNSAP	Radio Network Subsystem Application Part
RNTI	Radio Network Temporary Identity
routing	the layer 3 operation performed at an IP node which decides which node a packet should be sent to next, in order eventually to reach the destination
routing area update	similar to a location update but in this case the mobile recognizes a change of routing area ID (RAI) and updates the SLR in the SGSN
RPLMN	Registered PLMN
RPOA	Recognized Private Operating Agency
RR	Radio Resource
RRC	Radio Resource Control
RRM	Radio Resource Management
RSCP	Received Signal Code Power
RSE	Radio System Entity
RSL	Radio Signalling Link
RSS	Radio Subsystem: RSS = MS + BSS
RSSI	Received Signal Strength Indicator
RSVP	Resource Reservation Protocol
RSZI	Regional Subscription Zone Identity
RT	Real Time
RTE	Remote Terminal Emulator
RTP	Real-time Transport Protocol

RU   Resource Unit
RX   Receive
RXLEV   Received signal Level
RXQUAL   Received signal Quality

S
SAAL   Signalling ATM Adaptation Layer
SACCH   Slow Associated Control Channel
SABM   Set Asynchronous Balanced Mode
SA-STP   Stand-Alone Signalling Transfer Point
SACCH   Slow Associated Control Channel
SAD   Source Address
SAP   Service Access Point
SAPI   SAP Indicator
SAR   Segmentation And Reassembly
SAT   SIM Application Toolkit
SB   Synchronization Burst
SC   Service Centre
SC   Service Code
SCCH   Synchronization Control Channel
SCCP   Signalling Connection Control Part
SCCPCH   Secondary Common Control Physical Channel
SCF   Service Capability Feature (VHE/OSA context)
SCF   Service Control Function (IN context)
SCH   Synchronization Channel
SCI   Subscriber Controlled Input
SCN   Sub-Channel Number
SCP   Service Control Point – the IN element which realizes services
                    such as number translation and prepaid card billing
SCR   Sustainable Cell Rate
SDCCH   Stand-alone Dedicated Control Channel – the channel that
                    enables communication between the MS and BTS during
                    registration and call set-up
SDH   Synchronous Digital Hierarchy.
SDL   Specification Description Language
SDMA   Space Division Multiple Access
SDSL   Symmetrical Digital Subscriber Line
SDT   SDL Development Tool
SDU   Service Data Unit
SE   Support Entity
SEF   Support Entity Function
SF   Spreading Factor
SFH   Slow Frequency Hopping
SFN   System Frame Number
SGSN   Serving GPRS Support Node
SGW   Signalling Gateway

SHCCH	Shared Channel Control Channel
SI	Screening Indicator
SI	Service Inter-working
SI	Supplementary Information
SIC	Service Implementation Capabilities
SID	Silence Indicator
SIM	Subscriber Identity Module
SIM toolkit	allows software to run on the SIM card to realize new services
SIP	Session Initiation Protocol
SIR	Signal-to-Interference Ratio
SIR	Sustained Information Rate
SLA	Service Level Agreement
SLPP	Subscriber LCS Privacy Profile
SLR	SGSN Location Register – a logical element in the SGSN that has similar functionality to the VLR
SLTM	Signalling Link Test Message
SMDS	Switched Multi-megabit Data Service
SME	Short Message Entity
SMG	Special Mobile Group
SMI	Structure of Management Information
SMLC	Serving Mobile Location Centre
SMS	Short Message Service
SMSCB	SMS Cell Broadcast
SMS-SC	SMS – Service Centre
Smt	Short message terminal
SMTP	Simple Mail Transfer Protocol
SN	Serving Network
SN	Subscriber Number
SNAP	Sub Network Access Protocol
SNDCP	Sub Network Dependent Convergence Protocol – the protocol that compresses IP packets and segments them into LLC frames (and vice versa)
SNMP	Simple Network Management Protocol
SoLSA	Support of Localized Service Area
SOP	Support of Optimal Routing – enables local call setup between mobiles roaming in a foreign country, i.e. the speech channel does not go via the HPLMN
SP	Service Provider
SP	Signalling Point
SP	Switching Point
SPC	Signalling Point Code
SPCK	Service Provider Control Key
SQN	Sequence Number
SRNC	Serving Radio Network Controller
SRNS	Serving Radio Network Subsystem

SRES	Signed Response – part of a triple used for authentication and can be thought of as an electronic signature for the MS
SS	Supplementary Service
SS	System Simulator
SS7	Signalling System number 7 – this is a standardized system for the exchange of signalling messages between network elements, e.g. in GSM between an MSC and databases such as a HLR
SSC	Secondary Synchronization Code
SSC	Supplementary Service Control string
SSCF	Service Specific Coordination Function
SSCF-NNI	SSCF – Network Node Interface
SSCOP	Service Specific Connection Oriented Protocol
SSCS	Service Specific Convergence Sublayer
SSDT	Site Selection Diversity Transmission
SSF	Service Switching Function
SSN	SubSystem Number
SSP (or mSSP)	(mobile) Services Switching Point – element in an MSC that routes certain numbers to the SCP in an IN platform
SSSAR	Service Specific Segmentation And Reassembly sub-layer
STC	Signalling Transport Converter
STDM	Synchronous Time-Division Multiplex
STM	Synchronous Transfer Mode
STMR	Side Tone Masking Rating
STP	Signalling Transfer Point
STTD	Space Time Transmit Diversity
SVC	Switched Virtual Circuit
SVN	Software Version Number
SW	Status Word

T

T	Timer
T	Transparent
T	Type only
T1	Committee T1 – North-American standardization committee
TA	Terminal Adapter
TA	Timing Advance – refers to a control parameter for MS transmissions. It defines *when* the MS can transmit in order for the time slot to arrive at the BTS at precisely the right time without overlapping other time slots
TAC	Type Approval Code
TACS	Total Access Communication System
TAF	Terminal Adaptation Function
TAI	Timing Advance Index (TAI) – specifies the PTCCH sub-channel which a GPRS MS should make use of to obtain its timing advance parameter

TBF	Temporary Block Flow
TBR	Technical Basis for Regulation
TC	Transaction Capabilities
TC	Transmission Convergence
TCE	Trans Coding Equipment – converts 13 kbps voice coding used on Um to 64 kbps voice coding used in ISDN networks
TCE frame	subdivision of normal 64 kbps ISDN timeslot into 4×16 kbp time slots on the Abis interface which have a one-to-one mapping with time slots on the air interface
TCH	Traffic Channel – GSM user channel defined by a frequency and a time slot number
TCI	Transceiver Control Interface
TCP	Transmission Control Protocol
TC-TR	Technical Committee Technical Report
TD-SCDMA	Time Division – Synchronised CDMA
TDD	Time Division Duplex – one time slot for the uplink and another time slot for the downlink
TDMA	Time Division Multiple Access – enables many users to access the same resource by sharing the resource in the time domain, e.g. eight subscribers can use one frequency on the GSM air interface (*see* 'time slot')
TDoc	Temporary Document
TE	Terminal Equipment
TEI	Terminal Equipment Identifier
TEID	Tunnel Endpoint Identifier
TF	Transport Format
TFA	Transfer Allowed
TFC	Transport Format Combination
TFCI	TFC Indicator
TFCS	TFC Set
TFI	Temporary Flow Identifier
TFI	Transport Format Indicator
TFP	Transfer Prohibited
TFS	Transport Format Set
TFT	Traffic Flow Template
TI	Transaction Identifier
time slot	the physical resource available to one user in a TDMA system, e.g. a time slot on the GSM air interface can carry 13 kbps of voice for one user
TLLI	Temporary Logical Link Identifier
TLS	Transport Layer Security
TLV	Type, Length and Value
TM	Telecom Management
TMF	Telecom Management Forum
TMN	Telecommunication Management Network

TMSI	Temporary MS Identity – an identifier that identifies an MS to its current VLR (instead of using the IMSI) and can be changed regularly to maintain security
TN	Termination Node
TN	Time slot Number
TO	Telecom Operations map
TOA	Time Of Arrival
TON	Type Of Number
TP	Third Party
TPC	Transmit Power Control
TPDU	Transfer Protocol Data Unit
TR	Technical Report
training sequence	a fixed pattern of bits sent in every radio burst for synchronization purposes, e.g. a normal burst has a 26-bit training sequence
transceiver	equipment that can both transmit and receive radio signals
TrCH	Transport Channel
Triple	set of three codes (keys) used for authentication and ciphering purposes – a triple is composed of RAND, SRES and Kc
TRX	Transceiver often used to refer to a GSM RF carrier
TS	Technical Specification
TS	Tele Service
TS	Time Slot, e.g. the eight subdivisions of a carrier frequency on Um
TS0	Time Slot zero – TS are numbered form 0 to 7 on Um. TS0 is reserved on one frequency for the BCCH and other signalling channels
TSC	Training Sequence Code
TSDI	Transceiver Speech and Data Interface
TSG	Technical Specification Group
TSTD	Time Switched Transmit Diversity
TTA	Telecommunications Technology Association – South-Korean standardization committee
TTC	Telecommunication Technology Committee – Japanese standardization committee
TTCN	Tree and Tabular Combined Notation
TTI	Transmission Timing Interval
TUP	Telephone User Part (SS7)
TV	Type and Value
TX	Transmit
TXPWR	Transmit Power

U	
UARFCN	UTRA Absolute Radio Frequency Channel Number
UARFN	UTRA Absolute Radio Frequency Number
UART	Universal Asynchronous Receiver and Transmitter

UBR	Unspecified Bit Rate
UCR	UMTS Core Release
UDD	Unconstrained Delay Data
UDI	Unrestricted Digital Information
UDP	User Datagram Protocol
UDUB	User Determined User Busy
UE	User Equipment – the UMTS end device
UI	Unrestricted Information
UI	User Interface
UIC	Union Internationale des Chemins de Fer
UICC	Universal Integrated Circuit Card
UL	Uplink, i.e. from MS to BTS
Um	standard abbreviation for the air interface – often taken to mean 'Unlimited mobility'
UM	Unacknowledged Mode
UML	Unified Modelling Language
UMS	User Mobility Server
UMSC	UMTS MSC
UMTS	Universal Mobile Telecommunication System
UNI	User–Network Interface
UP	User Plane
UPC	User Parameter Control
UPCMI	Uniform PCM Interface
UPD	Up-to-date
UPT	Universal Personal Telecommunication
URA	User Registration Area
URA	UTRAN Registration Area
URAN	UMTS Radio Access Network
URI	Uniform Resource Identifier
URL	Uniform Resource Locator
USB	Universal Serial Bus
USC	UE Service Capabilities
USCH	Uplink Shared Channel
USF	Uplink State Flag
USIM	UMTS SIM
USSD	Unstructured Supplementary Service Data
UT	Universal Time
UTRA	UMTS Terrestrial Radio Access
UTRAN	UTRA Network
UUI	User-to-User Information
UUS	User-to-User Signalling – enables SMS to be transmitted during a voice call

V

V	Value only
VA	Voice Activity factor

VAD	Voice Activity Detection
VAP	Videotext Access Point
VASP	Value Added Service Provider
VBR-nrt	Variable Bit Rate – non-real time
VBR-rt	Variable Bit Rate – real time
VBS	Voice Broadcast Service
VC	Virtual Channel
VC	Virtual Circuit
VC	Virtual Connection – in ATM can be permanent (PVC) or switched (SVC)
VCI	Virtual Channel Identifier
VGC	Voice Group Call
VGCS	Voice Group Call Service
VHE	Virtual Home Environment
VLR	Visitor Location Register
VMSC	the visited MSC of a subscriber is the one the subscriber is currently using, either for a call or for mobility management (MM)
VoIP	Voice-Over IP
VP	Virtual Path
VPI	Virtual Path Identifier
VPLMN	Visited PLMN
VPN	Virtual Private Network
VSC	Videotext Service Centre

W

WAE	Wireless Application Environment
WAN	Wide Area Network
WAP	Wireless Application Protocol
WBEM	Web Based Enterprise Management
WCDMA	Wideband CDMA
WDMA	Wavelength Division Multiple Access
WDP	Wireless Datagram Protocol
WIN	Wireless Intelligent Network
WINEGLASS	Wireless IP NEtwork as a Generic platform for Location Aware Service Support
WG	Working Group
WLAN	Wireless LAN
WML	Wireless Mark-up Language
WPA	Wrong Password Attempts (counter)
WS	Work Station
WSP	Wireless Session Protocol
WTA	Wireless Telephony Applications
WTAI	WTA Interface
WTLS	Wireless Transport Layer Security
WTP	Wireless Transaction Protocol

WTX	Waiting Time Extension
WWT	Work Waiting Time
WWW	World Wide Web

X
| XID | eXchange Identifier |
| XRES | Expected user Response |

Y
| YATP | Yet Another Tunnelling Protocol |

Z
| ZC | Zone Code |

Index

GPRS Networks.   G. Sanders, L. Thorens, M. Reisky, O. Rulik, S. Deylitz
© 2003 John Wiley & Sons, Ltd.   ISBN: 0-470-85317-4